Issues in DAM GROUTING

Proceedings of the session sponsored
by the Geotechnical Engineering
Division of the American Society of Civil Engineers
in conjunction with the ASCE
Convention in Denver, Colorado

April 30, 1985

Edited by Wallace Hayward Baker

Published by the
American Society of Civil Engineers
345 East 47th Street
New York, New York 10017-2398

The Society is not responsible for any statements
made or opinions expressed in its publications.

Copyright © 1985 by the American Society of Civil Engineers,
All Rights Reserved.
Library of Congress Catalog Card No.: 85-70700
ISBN 0-87262-461-7
Manufactured in the United States of America.

PREFACE

In January, 1982, the Grouting Committee of the ASCE Geotechnical Engineering Division sponsored an International Conference on "Grouting in Geotechnical Engineering," in New Orleans, Louisiana. Since then, the Grouting Committee has focused on defining and iluminating the issues and directions set by the presenters and discussors at the New Orleans Conference.

Although perhaps the largest number of individual grouting projects do not involve dams, by far the largest volume of grout used is for grouting of new or existing dams. This volume presents eleven papers related to various issues in dam grouting.

Four papers focus on material properties related to grout penetrability and quality, particularly in relation to cement grout consistency. Three papers discuss the use of micro-computers to monitor and evaluate the dam grouting process. Three case histories discuss special dam grouting applications, including the use of asphalt grouts, micro-fine cement grouts, and chemical grouts to develop cut-off curtains below dams. Finally, the concept and preliminary test results of compaction grouting for earth dam foundation embankment densification is illustrated.

Each of the papers included in the Proceedings has received at least one positive peer view and has been accepted for publication by the Proceedings editor. All papers are eligible for discussion in the Geotechnical Engineering Journal and are eligible for ASCE awards.

It is hoped that the papers presented herein and the associated symposium discussion will serve to highlight and move towards resolution of several of the current important issues related to grouting of dams.

Wallace Hayward Baker

CONTENTS

I—MATERIALS

Permanence of Chemically Grouted Sands
    Raymond J. Krizek and Mark Madden 1
Grout Penetrability
    R. H. Karol 27
Cement Grouting: Water Minimizing Practices
    A. Clive Houlsby 34
Grout Slurries—Thick or Thin?
    Don U. Deere and Giovanni Lombardi 156

II—OPERATIONS

Use of Asphalt in Treatment of Dam Foundation Leakage—Stewartville Dam
    Boro Lukajic, Grant Smith, and John Deans 76
Microfine Cement Used to Grout Decomposed Granite Foundation Rock of Kanedaira Dam
    Mitsuhiro Nagawo .... *
Chemical Grout Curtains at Ox Mountain Dams
    Edward D. Graf, Daniel J. Rhoades, and Kenneth L. Faught 92
Embankment Foundation Densification by Compaction Grouting
    Wallace Hayward Baker 104
Computer Applications in Grouting
    Michael Demming, James L. Rogers, and Alex Tula 123
Computer-Assisted Grouting Evaluation Systems
    Leland F. Grant 132
Acoustic Emission Monitoring of Grout Movement
    Robert M. Koerner, Richard N. Sands, and James D. Leaird 149

Subject Index 165

Author Index 167

PERMANENCE OF CHEMICALLY GROUTED SANDS

Raymond J. Krizek[1], M. ASCE, and Mark Madden[2]

ABSTRACT -- The effectiveness of a grouting program depends not only on the complete permeation of the desired deposit with grout but also on the permanence of the grout when exposed to insitu conditions. Evaluated herein is the behavior of several chemically grouted soils when subjected to constant gradients as high as 100 for periods up to almost two years. The test specimens were prepared with one of three uniformly graded sands or a fine gravel and injected with one of five silicate or two polyacrylic grouts. The polyacrylic grouts showed no signs of deterioration with time and were unaffected by the applied gradient. The silicate-based sodium aluminate grout exhibited slight increases in permeability during the initial stages of the test, but the permeability stabilized after several weeks and the grout appeared to be relatively intact upon completion of the tests. The remaining silicate-based grouts showed large increases in permeability during the first weeks of testing, and the magnitude and rate of these increases were influenced by the curing time prior to testing and the magnitude of the applied gradient. Specimens cured for short times generally exhibited complete elutriation of the grout due to inadequate development of the grout strength. Older specimens displayed a gradual increase in permeability, but complete elutriation was rarely attained; the permeability increases in these cases appear to be due to erosion through voids in the grout caused by syneresis. The gradient influences only the rate at which the permeability increases and not its long-term value, which was generally observed to be one to two orders of magnitude less than that of the ungrouted soil, and independent of the grain size and density of the soil, and, for the most part, the proportions of components in the grout mixes. The small amount of grout that remained in these silicate grouted specimens after completion of the tests was found to be concentrated at the contact points between the soil grains.

INTRODUCTION

Two factors are largely responsible for determining the effectiveness of any grouting program; these are the thoroughness with which the grout permeates the soil deposit and the permanence of the grout once in place. The degree to which the grout permeates the soil deposit is strongly influenced by the viscosity and gel time of the grout and the void size distribution in the soil deposit, but the permanence of the grout is controlled by the strength of the gel and its durability when exposed to prevailing site conditions. Thus, a poorly selected grout may readily permeate a given soil deposit, but result in only a temporary improvement.

---

[1]Professor and Chairman, Department of Civil Engineering, Northwestern University, Evanston, Illinois
[2]Geotechnical Engineer, J. M. Lambe and Associates, Anchorage, Alaska

OBJECTIVES

This experimental program was designed to investigate the problem of grout permanence. The specific objectives were to (a) evaluate the time-dependent permeability of various grouted soil masses when subjected to a continuous flow of water under a high gradient for extended periods and (b) characterize the effect of various factors, both external and internal, on the long-term permanence of the grout.

FACTORS CONTROLLING EFFECTIVENESS OF GROUTING

Among the factors that control the effectiveness of a grouting program are the strength, durability, erosion, and dissolution of the grout and the pore size distribution of the soil. As used here, "strength" refers to the internal cohesion of the pure grout. Tallard and Caron (1977) suggested that a grout of inadequate strength would break down under pore water pressures caused by an applied gradient, with the result being a displacement of the grout and a sharp increase in the permeability. However, Cambefort and Caron (1957), using a sodium aluminate grout, and later Cambefort (1964) showed that grouts with very low gel strengths were able to withstand high pore water pressures. Obviously, other characteristics of the grouts are needed to reconcile these divergent conclusions.

Durability is the ability of a grout to withstand external influences without deteriorating or showing a significant reduction in mechanical properties. For example, a waterproofing grout must not be susceptible to dissolution or erosion in the presence of flowing water. Erosion is characterized by a gradual increase in permeability due to the slow elutriation of grout by flowing water; this behavior has been observed in silicate grouts by Cambefort and Caron (1957), Einstein and Schnitter (1970), Hurley and Thornburn (1972), and others. Einstein and Schnitter (1970) monitored the erosion process by periodic chemical analysis of the effluent from constant flow tests, and the concentration of silicate in the effluent at any given time was found to correlate with changes in the permeability. They concluded that the grout content within a soil tends toward a state of equilibrium under a given gradient, as indicated by the asymptotic leaching rates observed for all samples; in addition, the degree of grout erosion was determined to be a function of the average void size in the soil matrix and the magnitude of the applied gradient. Similar results reported by Rhone-Poulenc (1975) were found to closely parallel the rates of syneresis for the grout. In tests by both Karol (1957) and Clark (1982), no leaching or erosion was observed in specimens injected with polyacrylic grouts; since measured permeabilities were on the order of 10^{-8} cm/sec or less, this may be attributed to the low flow rates passing through the specimens.

Dissolution is the reversion of some or all of a solid gel to its liquid phase. Rhone-Poulenc (1975) attributed the dissolution of silica gels to excess concentrations of non-neutralized soda in the liquid phase of the gel which attacks the polysilicic acid in the silicate chains. The amount of soda is dependent on the proportion of sodium silicate in the grout, and the degree to which the soda is neutralized is a function of the proportion of reagent in the grout mix and the ionic strength of the reagent solution. Christopher (1979) attempted to cure specimens of silicate grouted sand in a water bath and found that those in which the gel was less than one hour old disintegrated immediately upon immersion. Although specimens aged for longer periods showed some deterioration, especially around the periphery, they were basically intact upon removal from the bath. This suggests that either the polymerized silicate becomes less susceptible to attack by non-neutralized soda with age or that the concentration of non-neutralized soda is decreased with time by neutralization with other ions in the liquid phase of the gel or by expulsion through syneresis. Studies by Karol (1957) and Clark (1982) found no similar deterioration of polyacrylic grouts when exposed to strong acids, bases, and solvents, in addition to static and flowing water; hence, the gels formed from polymerization of these polyacrylic compounds are apparently quite insoluble.

PREVIOUS STUDIES

In general, previous studies addressing this problem were designed to identify the most promising grouts for use under specific conditions; although the intent of the resulting test programs was of an endemic nature, important qualitative conclusions can be deduced. In the study by Einstein and Schnitter (1970), specimens injected with a sodium aluminate grout manifested a permeability increase of about one order of magnitude during the first several weeks, and this was followed by a more gradual decrease to a permeability similar to that measured at the start of the test. These observations agree with those of Cambefort and Caron (1957). In both studies, the maximum permeability was on the order of 10^{-5} cm/sec at gradients of up to 250. Einstein and Schnitter (1970) attributed the permeability increase early in the tests to contractions in the gel and to the dislodgement of some grout caused by the applied gradient and the subsequent permeability decrease to the migration of the dislodged material through the specimen until it blocked other voids.

In permeability tests with Hardener 600, Rhone-Poulenc (1975) reported a similar behavior pattern, except that the increase during the initial stages of the test was several orders of magnitude and resulted in permeabilities only one to two orders of magnitude less than that of the ungrouted sand. After reaching this peak, the permeability decreased slowly throughout the remainder of the test. Hurley and Thornburn (1972) reported similar results, except that, following the sharp initial increase, the permeability continued to increase gradually. Specimens injected with an ethyl acetate-formamide grout by Benltayf (1980) showed large permeability increases when cured for less than four days. The grout was displaced as a viscous fluid and complete elutriation occurred within several hours, resulting in a permeability only slightly less than that of the ungrouted sand. Specimens cured for longer periods exhibited a nearly constant permeability in the range of approximately two orders of magnitude less than that of the ungrouted sand. The behavior of the specimens cured for short periods is consistent with predictions by Tallard and Caron (1977), but the lack of variation in the permeability during the early stages of tests on older specimens is in contrast with observations by other researchers working with silicate grouts. This difference may be due to the low gradients and short test durations used by Benltayf (1980). However, in three tests of extended duration the permeability of the specimens remained essentially constant until near the end of the tests at approximately 50 days, at which time the permeability began to increase rapidly, thus indicating that the time required for the initial permeability increase to occur in silicate grouts may be a function of both the age of the grout and the applied gradient.

Davidson and Perez (1982) tested the effectiveness of several grouts in situ. Using borehole falling head permeability tests, they found a decrease in the permeability of an alluvial sand deposit to be about two orders of magnitude due to the injection of a sodium aluminate grout. Laboratory tests on sand from the site mixed with similar grouts showed permeability reductions of four to five orders of magnitude. This difference between laboratory and field results was attributed to the difference in the methods by which the grout was applied (mixing in the lab as opposed to injection in the field) and to incomplete permeation of the grout in the field deposit.

MATERIALS TESTED

Seven chemical grouts of two major types (five silicate and two polyacrylic) were tested; these represent more than 90% of all chemical grouts presently used in engineering practice (Tallard and Caron, 1977). The soils included in this study were three grades of standard Ottawa sand and a fine gravel. The water used as the permeant (but not necessarily as a component in the grout mixtures) was tap water from the Evanston, Illinois municipal water supply system.

Silicate Based Grouts

Five different reagents were used with sodium silicate to form grout solutions. Two were standard laboratory compounds, namely, sodium aluminate ($NaAlO_2$) and a mixture of ethyl acetate ($CH_3COOC_2H_5$) and formamide ($HCONH_2$). The remaining three were commercially marketed reagents known as Hardener 600, Terraset, and Geloc-4. The silicate for all of these grouts was supplied in the form of commercially available "N" grade sodium silicate manufactured by the Philadelphia Quartz Company; this solution consists of 28.9% SiO_2 and 9.8% Na_2O by weight, representing a weight ratio (SiO_2/Na_2O) of 3.22. At 20°C, "N" grade sodium silicate has a viscosity of 180 cp, a specific gravity of 1.38, and a pH of 11.3.

<u>Ethyl Acetate and Formamide</u> -- The gel formed by combining a dilute solution of ethyl acetate and formamide with sodium silicate takes advantage of the desirable characteristics produced by both reagents. Ethyl acetate is only partially soluble in water, but reacts very rapidly with sodium silicate to form a high strength gel. On the other hand, formamide is completely soluble in water and reacts very slowly with the silicate to form a homogeneous but weak gel. Combining ethyl acetate and formamide in proper proportions will produce a strong grout whose gel time can be reasonably well controlled. In

this study equal portions (by volume) of the two reagents were mixed together until they were completely dissolved, and this solution was then diluted with a specified volume of water, mixed thoroughly, combined with the sodium silicate, and again mixed thoroughly. Important physical characteristics of the ungelled solution were a pH of 11.2, a specific gravity of 1.20, and a viscosity of 5 cp at 20°C.

Sodium Aluminate -- The gelation process resulting from the combination of dilute sodium aluminate and sodium silicate follows a pattern similar to that of pure silicate, except that both orthosilicic and alumino-silicic acid ions are involved in the reaction. The end product is a nearly amorphous gel consisting of short alumino-silicate chains. The grout exhibits a low strength that is characteristic of its poorly defined structure, but the alumino-silicate molecules have the capacity to store large unincorporated molecules or ions, such as water or hydrated sodium, within their matrix (Merrill and Spencer, 1950). Plank and Drake (1947) and Plank (1947) conducted extensive tests with alumino-silicate gels and concluded that leaching of these interstitial ions from the gel by extensive washing was minimal. The gel time of the grout mixture is dependent on the weight ratio of aluminate to silicate (Al_2O_3/SiO_2) in solution; however, once gelation has occurred, the gel displays very little syneresis (this is believed to be due to its amorphous structure). The sodium aluminate used in this investigation contained 19.9% aluminate by weight and was marketed as #2 by the Nalco Chemical Company. The mixing procedure was to divide the specified volume of water into two beakers into which either the sodium aluminate or sodium silicate was dissolved; the two solutions were then combined and thoroughly mixed to form the grout. It was necessary to use distilled water in this mixture because small amounts of an insoluble aluminum precipitate formed whenever the sodium aluminate came in contact with tap water. The grout mixture had a pH of 11.0, a specific gravity of 1.12, and a viscosity of 3 cp at 20°C.

Commercial Reagents -- Three commercially available reagents -- Hardener 600 (Rhone-Poulenc of Paris, France), Terraset (Celtite Incorporated of Cleveland, Ohio), and Geloc-4 (Hayward Baker Company of Odenton, Maryland) -- were used. The composition of each is proprietary information, but all are described as compound diacid esters, probably of a similar chemical nature (Rhone-Poulenc, 1975; Celtite, 1980; Baker, 1982). The gelation process is believed to be similar to that of the ethyl acetate and formamide mixture, thereby resulting in a similar end product. In all cases the gel time of the mix is controlled by the concentration of silicate and hardener in solution; Terraset also has available an accelerator in the event that very rapid gel times are desired. The typical mixing method was to dissolve the hardener in a specified volume of water and then combine this solution with sodium silicate. Hardener 600 and Geloc-4 were not readily soluble in water, and they required high speed mixing to achieve a reasonable dispersion of the reagent; incomplete dispersion resulted in poor control of gel time and a non-homogeneous gel. Typically, the fluid grouts had a pH of 11, a specific gravity of 1.20, and a viscosity of 5 cp at 20°C.

Polyacrylic Based Grouts

Two polyacrylic grouts -- AM-9 and AC-400 -- were tested in this study. Polyacrylic grouts are typically a collection of water soluble monomers and polymers which, when mixed with a catalyst system, polymerize to an insoluble elastic gel. The catalyst system consists of an initiating compound and an accelerator, and the polymerization is a two-step process. Upon mixing, the initiator and accelerator combine to form a reaction product, the production rate of which is controlled by the concentration of the catalyst in solution. This process continues until a critical concentration of reaction product is reached, causing a rapid exothermic polymerization of the monomers and polymers in solution. The polymerization reaction is complete in several minutes and results in an elastic gel shown to be insoluble in either strong acids or alkalines (Karol, 1957; Clark, 1982). Unlike silicate grouts, which show a gradual increase in viscosity prior to gelation, the viscosity of the acrylic solutions remains essentially constant until the gel time is reached. Because gelation involves complete polymerization, the resulting gel, which consists of 80% to 90% water entrapped within the polymer matrix, undergoes no syneresis. The common mixing procedure employed for both of these grouts was to make two solutions of equal volume and then combine them to form the grout; one solution contained the the polymers (or monomers) and accelerator, and the other contained the initiator. Because there are no chemicals of high ionic strength and all constituents are water soluble, the physical properties of the resulting mixtures were similar to those of water; typical values are a pH between 7 and 9, a specific gravity ranging from 1.00 to 1.05, and a viscosity of less than 3 cp at 20°C.

AM-9 -- The chemical grout AM-9 was produced by the American Cyanamide Company of Wayne, New Jersey until 1978, at which time it was removed from the U.S. market due to toxicity problems. According to Karol (1957), AM-9 is a mixture of acrylamide and N,N'-methylenebisacrylamide, and the catalyst system includes an initiator (ammonium persulfate (AP)) and an accelerator (either nitrilotrispropionamide (NTP), β-dimethylaminopropionamide (DAP), diethylaminopropionamide (DEAPN), or dimethylaminopropionamide (DMAPN)).

AC-400 -- Designed as a replacement for AM-9 (Clark, 1982), AC-400 is an acrylate polymer grout produced by the Geochemical Corporation of Ridgewood, New Jersey; it contains a mixture of acrylate monomers cross-linked with a small amount of methylenebisacrylamide (MBA), and the catalyst system consists of an initiator (ammonium persulfate (AP)) and an accelerator (triethylamine (TEA)). The gel time can be controlled by varying the concentration of AP and TEA in solution; longer gel times can be achieved by the addition of small amounts of potassium ferricyanide (KFe).

Soils

The index properties and the coefficient of permeability, k_o, for each of the four soils are summarized in Table 1.

Table 1. Descriptions of Soils Tested

	Ottawa 20-30	Ottawa Bond	Ottawa F-140	Fine Gravel
G_s	2.67	2.67	2.67	2.64
D_{10} (mm)	0.59	0.16	0.053	4.76
C_u	1.00	1.75	1.98	1.10
e_{min}	0.49	0.48	0.47	0.38
e_{max}	0.67	0.77	0.78	0.72
Δe	0.18	0.29	0.31	0.34
k_o (cm/sec)	0.03	0.01	0.003	0.05

SCOPE OF EXPERIMENTAL PROGRAM

This study was divided into two phases; the first phase was directed toward evaluating the effects of different soils and grouts on the long-term permeability of the grouted soil, and the second phase was designed to assess the influence of curing time on grout permanence. In Phase 1 more than thirty specimens were prepared using combinations (although the permutation test matrix was not complete) of five different grouts (not including Geloc-4 and sodium aluminate), eight different grout mixtures, and four different soils in a dense state; the eight different grout mixtures consisted of one mixture each for the Terraset, AM-9, and AC-400 grouts and the five mixtures (see Table 2) that were used to assess the effects of increased silicate content in the Hardener 600 and ethyl acetate-formamide grouts, as well as the percentage of non-neutralized soda in the former. These specimens were cured for seven days and then tested at a constant gradient of 100 for a period of 16 to 21 months or until the permeability approached that of the ungrouted sand, k_o. In Phase 2 Ottawa 20-30 sand in a loose state was used to prepare a series of eight specimens for each of seven grout types; after curing for various periods (0, 10, and 30 minutes, 1 and 6 hours, 1, 3, and 7 days),

Table 2. Mixtures of Hardener 600 and Ethyl Acetate-Formamide Grouts

	Concentration (by % of volume)			Neutralization
	Silicate	Reagent	Water	(%)
Hardener 600	50 50 60	8 12 10.5	42 38 29.5	50 75 50
Ethyl Acetate and Formamide	50 50	11.5 13.5	38.5 26.5	100 100

these specimens were subjected to a gradient of either 50 or 100 for three months or until the permeability reached k_p. Due to certain common features, some data from Phase 2 can be used to complement that from Phase 1 when analyzing the response patterns.

Each specimen was identified by up to three sequences of letters or numbers. The first sequence was a two letter code identifying the grout type and sand type, respectively; the letter codes are given in Table 3. In Phase I the second and third sequences, which indicate the silicate concentration and percentage of neutralization, respectively, were employed only for specimens grouted with the ethyl acetate-formamide and Hardener 600 grouts. In Phase 2 the second alpha-numeric sequence designated the curing time in minutes (M), hours (H), or days (D), and the third sequence was used only to designate duplicate specimens.

Table 3. Specimen Identification Legend

```
                    Grouts
    F = Ethyl Acetate and Formamide
    M = AM-9      S = Sodium Aluminate
    C = AC-400    H = Hardener 600
    G = Geloc-4   T = Terraset

                    Soils
          C = Ottawa 20-30 Sand
          M = Ottawa Bond Sand
          F = Ottawa Bond Sand
          G = Fine Gravel
```

EXPERIMENTAL PROCEDURES

The experimental procedures employed in this test program can be conveniently divided into four phases -- specimen preparation, grout injection, curing, and testing.

Specimen Preparation

The permeameters (see Figure 1) consisted of 12-inch lengths of 2-inch diameter, heavy-wall, glass tubing, sealed at each end with a rubber stopper; this unit was restrained between two lucite plates connected at each corner by threaded rods. A 0.375-inch diameter hole was bored into each rubber stopper to accommodate flow connections. After the specimen was grouted, a 0.5 inch layer of ungrouted

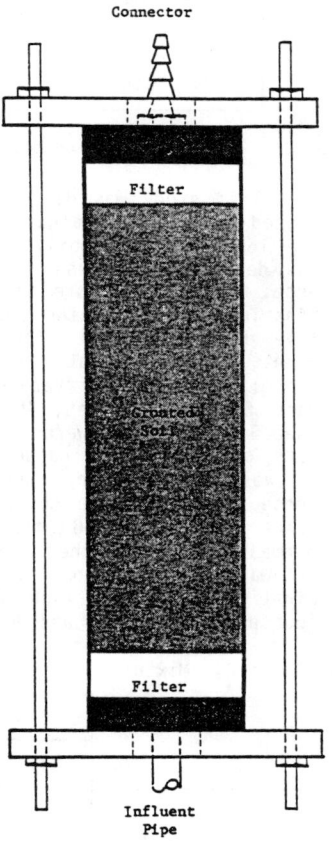

Figure 1. Schematic Diagram of Permeameter

Ottawa 20-30 sand was placed at each end; these filter layers, combined with the penetration of the rubber stoppers, resulted in an actual specimen length of approximately 10 inches.

Since both dense and loose specimens were prepared in the two phases, two different procedures were used for placement of the sand. In both cases sand was pluviated into the permeameter through a funnel positioned to maintain a constant height of drop of approximately 2 inches. Loose specimens were made by the use of pluviation alone, whereas dense specimens were constructed in four lifts and vibrated after placement of each lift with a small hand-held vibrator for a period of one minute at each of three points on the circumference of the specimen at intervals of 120°. The average relative densities of the specimens are shown in Table 4; only specimens with void ratios within ±0.03 of the values given in Table 4 were considered acceptable for testing.

Table 4. Average Densities of Specimens

	Condition	Void Ratio	Density (%)
Gravel	Medium	0.56	47.1
20-30 Sand	Dense	0.54	72.2
20-30 Sand	Loose	0.63	22.2
Bond Sand	Dense	0.56	72.4
F-140 Sand	Dense	0.50	90.3

Grout Injection

The grout injection process was designed to assure a uniform distribution of grout while causing the least disturbance in the sand. The injection system (see Figure 2) consisted of a pumping apparatus and a frame to accommodate the specimens during injection. The pumping apparatus included a regulated air supply, a grout supply tank, a control valve, an injection nozzle, and connecting hoses. The air pressure in the system was controlled by a regulator and pressure gage that was accurate to 0.1 psi. The system control valve was a three-way brass valve that permitted flow to be diverted to either the injection nozzle or the sample port, as well as complete flow stoppage; this arrangement allowed for sampling of the grout or termination of flow without disconnecting the injection or feed lines. The injection nozzle was a brass hose connector which was ground slightly to facilitate easier insertion and removal from the flexible tubing. All flexible tubing consisted of inert Tygon material with an outside diameter of 0.375 inches. The grout supply tank was a lucite cylinder with an approximate volume of 4500 cm^3 reinforced by metal screening. Pumping was accomplished by forcing grout out of the tank through a bottom hose by means of air pressure supplied through a top connection; the pumping rate was controlled by manipulating the air pressure, and the flow path of the grout was determined by the control valve.

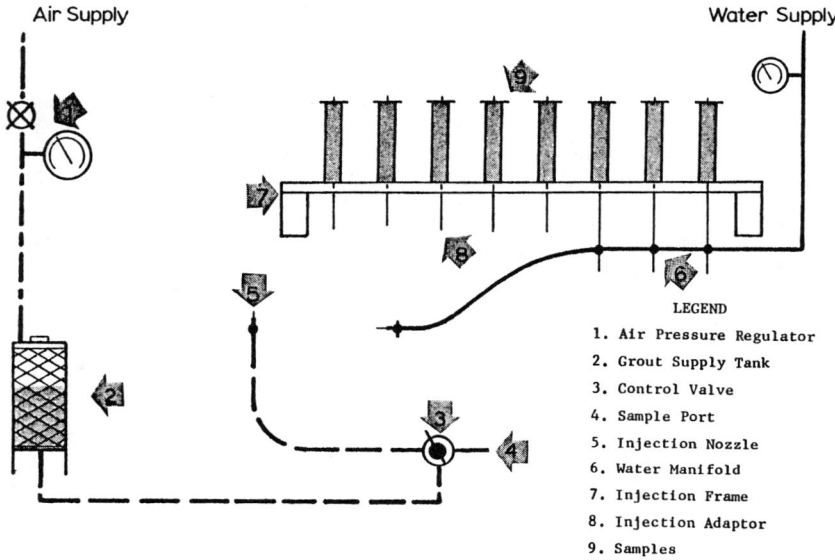

Figure 2. Schematic Lay-out of Grout Injection System

LEGEND
1. Air Pressure Regulator
2. Grout Supply Tank
3. Control Valve
4. Sample Port
5. Injection Nozzle
6. Water Manifold
7. Injection Frame
8. Injection Adaptor
9. Samples

The sample frame consisted of a 5-foot long wooden beam positioned about 6 inches above the table top and divided into eight bays, each consisting of a well for the base of the specimen and a restraint system. The specimens were connected to a manifold which allowed the injection of either water or grout by manipulating the valving arrangement. Initially water was flushed through each specimen, saturating the sand and allowing the permeability of the ungrouted specimen, k_o, to be measured. In preparation for grout injection batches of 750 cm^3 (approximately three times the void volume in an average specimen) were mixed; the gel times in Phase 1 were usually on the order of 30 minutes, whereas in Phase 2 gel times ranged from 10 to 15 minutes. Upon mixing, the grout was immediately transferred to the grout supply tank, which was then pressurized to within 0.5 psi of the injection pressure given in Table 5; the injection pressure was determined to be that required to pump two void volumes of grout into the specimen within three minutes. Following complete saturation of all lines with grout, a sample of grout was taken and injection was begun.

During the injection process, two tests were performed to estimate the extent of grout saturation. First, samples of the effluent were taken at various times and the pH of these samples was compared with those of the pure grout and the water supply. And second, all effluent was collected and, with a knowledge of the volume and specific gravity of the effluent and the specific gravity of the pure grout, the volume of water in the effluent was estimated and compared

Table 5. Grout Injection Pressures

Grout	Injection Pressure (psi)
Ethyl Acetate-Formamide	4
Hardener 600	8
Geloc-4	5
Terraset	5
Sodium Aluminate	3
AC-400	2

to the void volume of the specimen. These tests proved very sensitive to volume and weight measurements, especially for the polyacrylic grouts whose specific gravity and pH were very similar to those of the water supply. As a further check, no ungrouted pockets were observed in several specimens which were dismantled immediately after gelation. The injection process was terminated when the supply tank was nearly empty; a second sample of grout was taken; and the specimen was sealed immediately by closing all manifold valves. The pumping system was then purged and thoroughly flushed with water. The gel times of the grout samples taken at the beginning and end of injection were compared to the design gel time to ensure that the grout solution remained homogeneous throughout the injection process.

Curing

During the curing period, the specimens remained sealed. Specimens cured for less than six hours were left in the injection frame until ready to be tested, but specimens cured for longer periods were removed from the grouting frame several hours after injection and transferred to a high humidity environment for storage.

Testing Procedures

The testing set-up (see Figure 3) consisted of a water supply and a distribution system. The supply system provided a continuous supply of tap water at a constant pressure of 40 ±5 psi and the distribution system consisted of brass tee fittings connected in series at 8-inch intervals along several parallel branches. Pressure was monitored across the system by gages installed at both ends. Flow was from bottom to top and a flexible hose was connected to the outlet end of each permeameter to carry the effluent to a disposal system. During the first hours of each test, detailed observations and permeability values were recorded, but readings became less frequent as the flow rate through the specimen stablilized. Although the high flow rates in many of the specimens probably rendered the use of Darcy's law inappropriate, the exact quantitative values of the permeability in these cases were not of primary interest. All tests were continued until either the prescribed term had elapsed or until the permeability of the specimen reached k_o. Upon termination, each specimen was dismantled and visual observations were recorded.

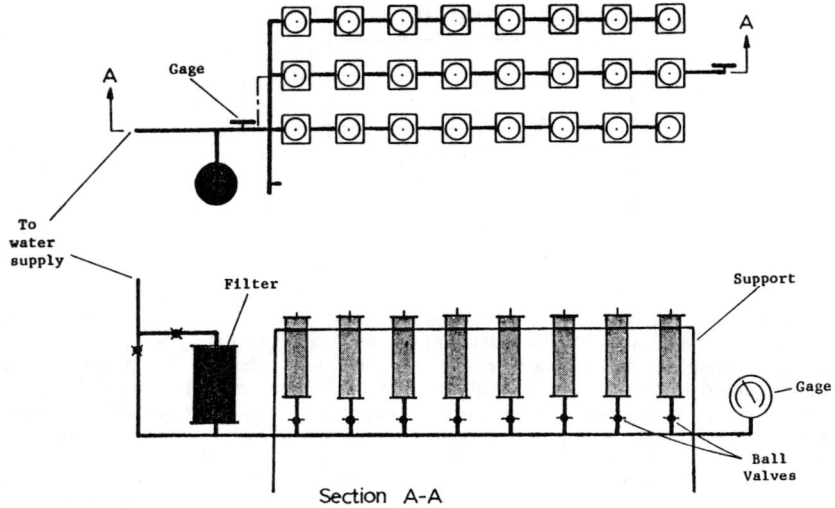

Figure 3. Schematic Lay-out of Permeability Test Apparatus

INTERPRETATION OF RESULTS

Data from approximately one hundred constant-head permeability tests indicate that any major changes in permeability that were going to occur did so during the first several weeks. After this time, the permeability of all specimens remained essentially constant throughout the duration of the tests. Based on these data and observations made upon dismantling the specimens, several factors have been identified which affect both the long-term and short-term permeability of grouted soils.

Short-Term Behavior

An examination of data from the initial stages of all tests showed that most variations in permeability occurred during the first several weeks. The factors found to have the most significant influence on this variation were the grout type, curing time, and applied gradient. Figure 4 shows three typical patterns into which all of the observed short-term behavior conforms. In Pattern 1 the initial permeability remains essentially constant until a dramatic increase takes place, with values changing as much as four orders of magnitude in several minutes. Following this increase, the permeability remains relatively constant for the remainder of the test. Pattern 2 is characterized by a gradual increase in permeability until some peak value is reached, after which the permeability remains constant until the test is terminated; this gradual increase covers two to three orders of magnitude over a period lasting from several hours to several weeks. Specimens following Pattern 3 showed no change in permeability throughout the test.

Grout Type -- The type of grout injected into the specimen strongly influenced the behavioral pattern. All specimens injected with Terraset followed Pattern 1; in these specimens the initial period during which the permeability remained constant lasted from zero to three hours, after which jumps of three to four orders of magnitude were observed in a period of several minutes. Specimens grouted with sodium aluminate generally followed Pattern 2; initial permeabilities of 10^{-7} cm/sec were observed upon application of the gradient, but the permeability increased gradually and exhibited increases of one to three orders of magnitude over periods of 20 to 30 days. The general behavior of specimens injected with AM-9 and AC-400 is exemplified by Pattern 3; in these specimens the flow rate was less than 1 cm^3/day, representing a permeability on the order of 10^{-9} cm/sec. The remaining grouts followed either Pattern 1 or 2, depending on the time the grout was allowed to cure prior to testing. However, there were several exceptions to the generalizations discussed above, and these will be explained later.

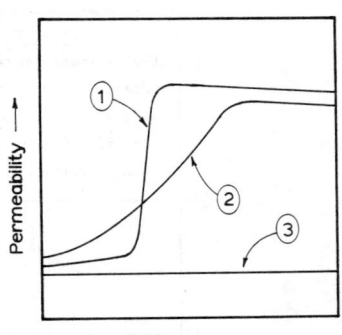

Figure 4. Patterns of Short-Term Permeability Behavior

Curing Time -- The effects of curing time may be related to the increase in gel strength with time and syneresis. Tallard and Caron (1977) suggest that grouts of inadequate strength to withstand pore pressures due to an applied gradient will become dislodged, resulting in a sharp increase in permeability. Since strength has been shown to increase with grout age (Christopher, 1979), it would be expected that older or stronger grouts would be less likely to follow Pattern 1. On the other hand, shrinkage of the gel due to syneresis will result in the formation of a series of interconnected cavities within the grouted soil matrix (Rhone-Poulenc, 1975), and, since the effects of syneresis are known to increase with grout age, it would be expected that older specimens and specimens injected with grouts that exhibit greater amounts of syneresis would be more prone to erosion because the larger cavities increase the flow rate. This effect would most likely be manifested as a reduction in the initial stage of Pattern 1 or a reduction in the period over which the permeability increases in Pattern 2. To examine the relative (these results are considered to be "relative" because syneresis is a complicated phenomenon that is believed to depend, among other factors, on the dimensions of the grout sample) amount of syneresis among the silicate grouts used in this study, 100 cm^3 samples of each grout were placed in glass jars of similar dimensions; the rate of syneresis over a period of 21 days following gelation is illustrated in Figure 5.

All silicate grouted specimens in Phase 2 (excluding sodium aluminate) reached or exceeded a permeability of one order of magnitude less than the ungrouted permeability of the specimen (that is 0.1 k_o)

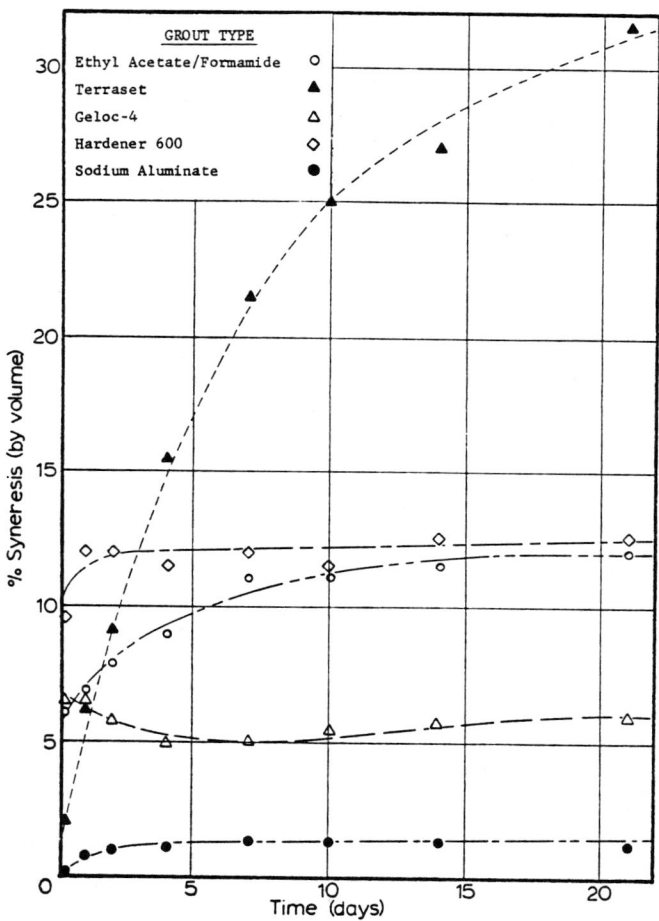

Figure 5. Syneresis Rates of Silicate Grouts

during the initial stages of each test, and the data in Figure 6 show that the time required to reach this value is consistent with the foregoing rationale. The most dramatic effects of syneresis is manifested by the Terraset specimens, where the time to reach 0.1 k_o ranged from less than one-half hour to a few hours and, after a brief curing period, decreased with increased curing time. Only two of the Terraset specimens in Phase 2 did not elutriate completely; these were the specimens cured for one and six hours. Specimens cured for longer than six hours are believed to have undergone such a large amount of syneresis that high flow rates eroded the remaining grout at a very rapid rate.

By comparison, the syneresis observed in samples of the other silicate grouts showed significant activity only during the first day or so. Christopher (1979) observed that specimens injected with ethyl acetate-formamide grout achieved their maximum strength after curing for about four days; this period coincides with the peak point on the ethyl acetate-formamide curve in Figure 6. The curves for Hardener 600 in both Figures 5 and 6 closely parallel those of the ethyl acetate-formamide grout, indicating similar strength and syneresis behavior. The curve for Geloc-4 in Figure 6 is generally higher than

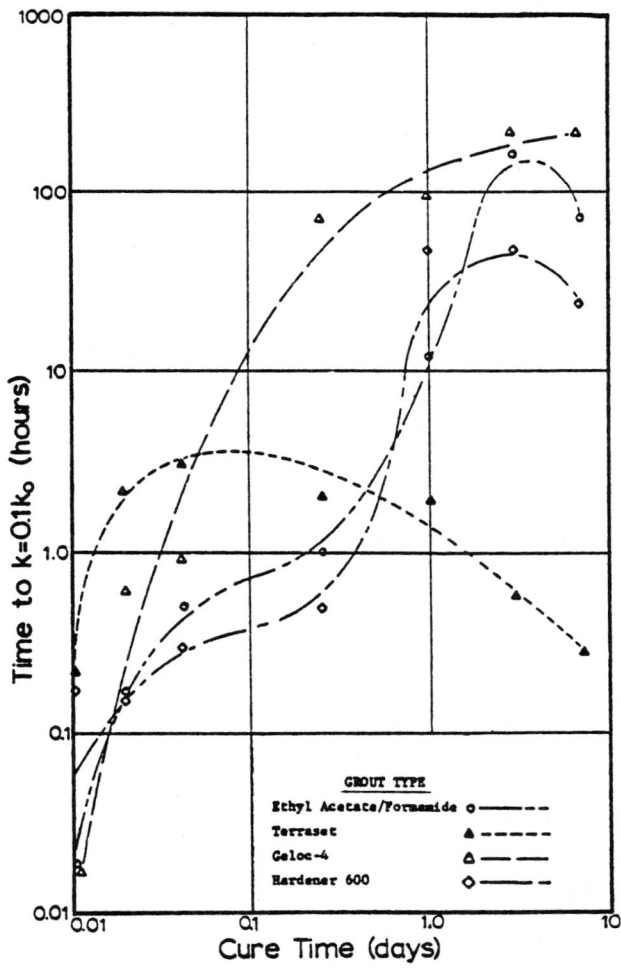

Figure 6. Effect of Curing Time on Elutriation Rate in Silicate Grouted Specimens

curves for the other grouts, and this is consistent with the small amounts of syneresis indicated in Figure 5.

The permeability of all specimens cured for less than one day (except the Geloc-4 specimen cured for six hours) followed Pattern 1, and during the permeability jump grout was observed to elutriate from the specimens as a viscous fluid. The permeability of the remaining specimens (except for those grouted with Terraset) followed Pattern 2, thus suggesting that grout elutriation was most likely due to erosion stemming from the effects of syneresis. Data from specimens grouted with the polyacrylic and sodium aluminate grouts are not plotted in Figure 6 because in only a few cases (namely, the polyacrylic specimens tested prior to gelation and the sodium aluminate specimens cured for less than thirty minutes) did the permeability ever reach $0.1k_0$ and in those cases the grout was elutriated immediately upon application of the gradient. In these grouts the strength of the gel is achieved very rapidly and little or no syneresis is observed. One phenomenon (for which no explanation is offered) is the trend depicted in Figure 7 toward higher maximum permeability values with increased curing times for sodium aluminate specimens; a similar trend was not observed for the other silicate grouts.

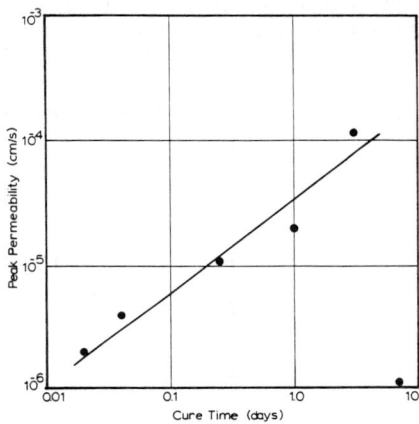

Figure 7. Peak Permeability versus Curing Time for Sodium Aluminate Specimens

Gradient -- Most specimens were tested under a gradient of 100; to assess the effect of gradient on permeability, four tests were conducted at a reduced gradient of 50. As seen in Figure 8, specimens tested at this reduced gradient followed patterns similar to their counterparts tested under a gradient of 100. The final permeability achieved under either gradient was similar, and the only significant difference attributed to the reduction in gradient was the time required for the permeability increase to occur. This observation is further supported by data reported by Benltayf (1980), where an increase in permeability in specimens injected with ethyl acetate-formamide grout did not occur until after 30 days under a gradient of about 10.

Long-Term Behavior

Upon completion of the short-term variations, the permeability of all specimens remained relatively constant throughout the remainder of the test. The permeability at which each sample stabilized was influenced by the grout type, average void size, and, to a lesser extent, grout mixture.

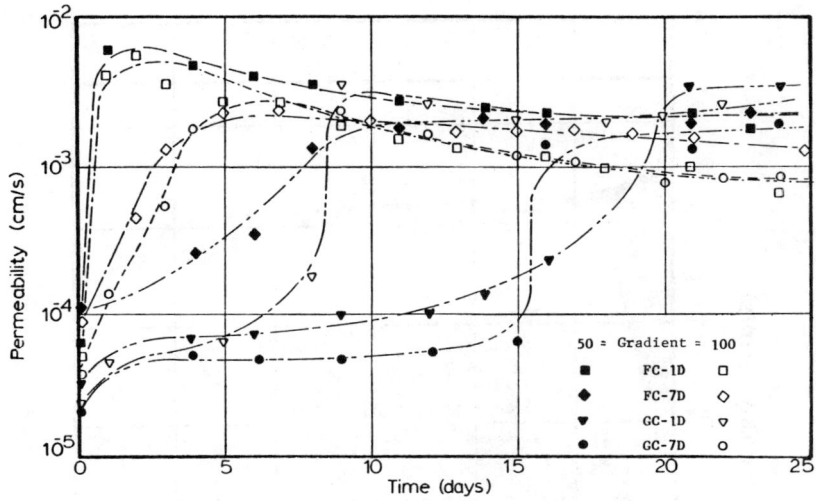

Figure 8. Effect of Gradient on Short-term Behavior of Silicate Grouted Specimens

Grout Type --The long-term permeability of all specimens can be grouped into three bands, depending upon the grout type. Figure 9 shows the results for all specimens prepared with Ottawa 20-30 sand.

Average Void Size -- Due largely to the different soils used in this test program, the average void size and associated ungrouted permeability (see Table 1) were important variables. Time-dependent variations in the permeability of the grouted specimens ranged from imperceptible differences for those injected with polyacrylic grouts to several orders of magnitude for those injected with silicate grouts. Figure 10 shows the variations in permeability for silicate grouted specimens; note that, regardless of grain size or density, the permeability of a grouted specimen was always one to two orders of magnitude less than that of the ungrouted specimen. Although sodium aluminate grout was tested in only one soil, it appears to be an exception to this observation. In the case of the gravel specimens, tests were conducted using AM-9, AC-400, and Terraset grouts cured for seven days only. In the gravel and coarse sand specimens injected with Terraset, the grout was elutriated immediately. The gravel specimen grouted with AC-400 exhibited a constant permeability of about 10^{-4} cm/sec throughout the test, but, upon dismantling this specimen, it was concluded that flow took place along the boundary. The gravel specimen grouted with AM-9 manifested a constant permeability of less than 10^{-9} cm/sec, which is similar to all other specimens with the same grout.

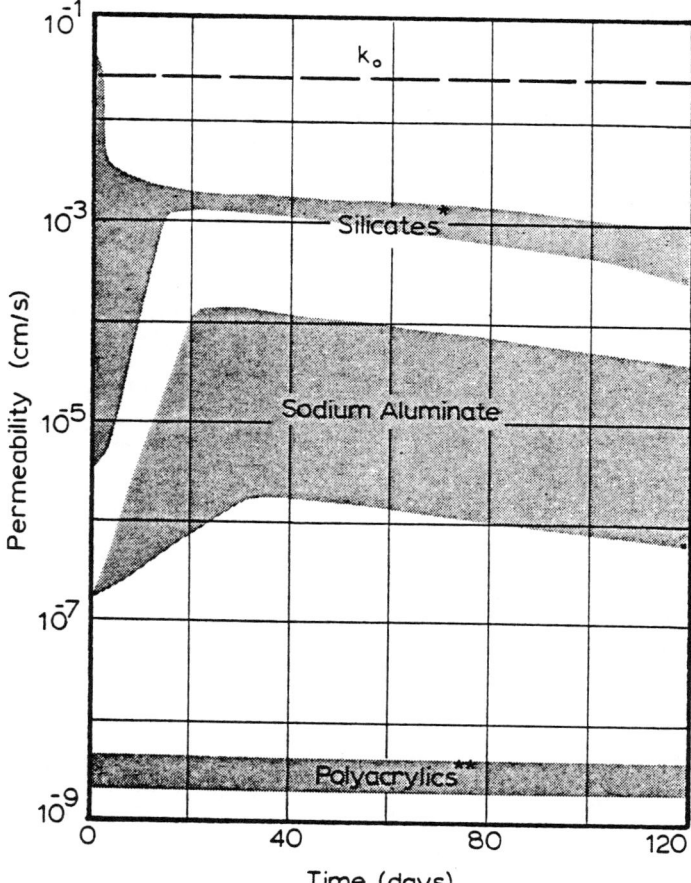

* Includes Hardener 600, Geloc-4, Terraset, and ethyl acetate/formamide grouts
** Includes AM-9 and AC-400

Figure 9. Permeabilities of Various Grouts

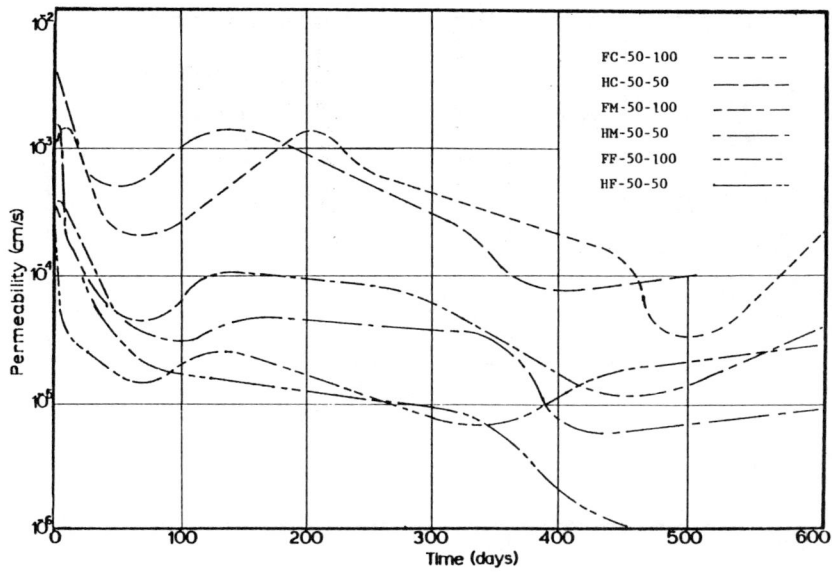

Figure 10. Effect of Grain Size on Permeability of Silicate Grouted Specimens

Grout Mix -- The silicate content and/or percentage of non-neutralized soda was varied in the Hardener 600 and ethyl acetate-formamide grouts. A decrease in permeability was expected as the former increased and the latter decreased, because either should result in a stronger gel. However, as seen in Figures 11 and 12, the variations were generally very small.

System Effects -- Throughout the test program, a slight decrease in permeability was noted in most tests conducted over longer periods. This decrease, observed to be approximately one order of magnitude per year, was originally believed to be due to changes in the grout structure with time; other researchers (Rhone-Poulenc, 1975; Cambefort and Caron, 1957) have observed similar behavior. However, evidence also exists to suggest that this decline in permeability might be due to problems with the test procedure. For example, the rate of decline was greatest in the specimens that experienced higher flow rates. In addition, an orange residue, believed to be iron compounds precipitated from the feed water, was found in most of the specimens dismantled at the end of Phase 1. This orange material was confined to the inlet filter and as a film on the face of the fine-grained and medium-grained specimens, but it appeared as a light stain throughout much of the length of the coarse-grained specimens. To examine the effects of this precipitated material, several medium-

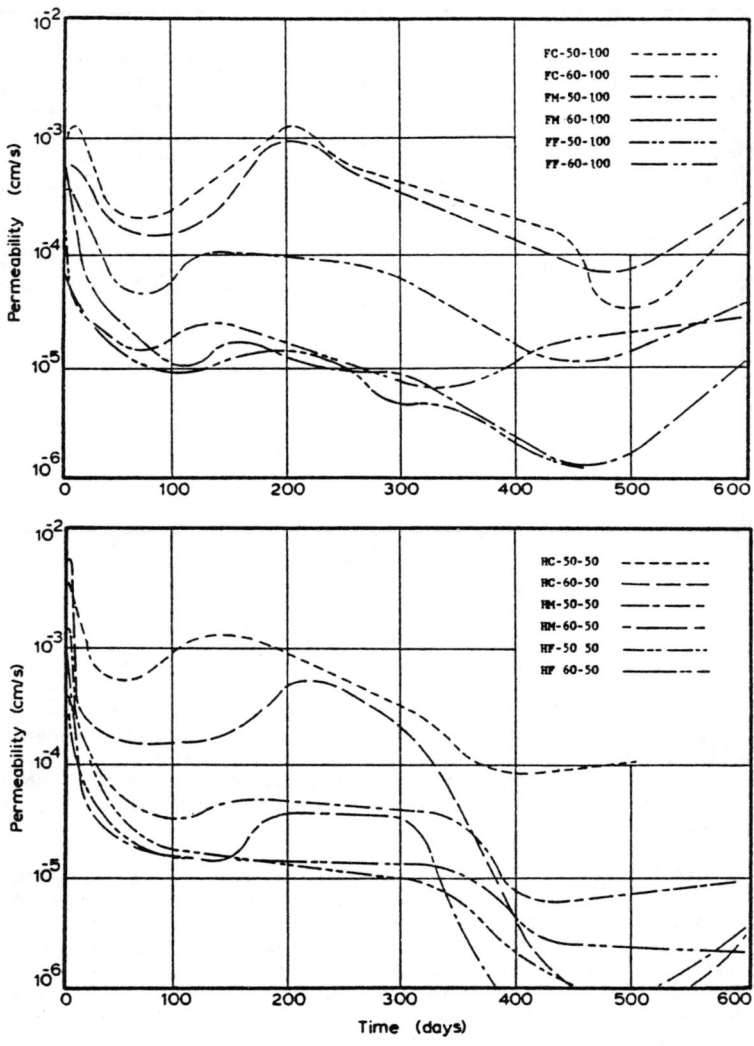

Figure 11. Effect of Silicate Concentration on Long-term Permeability

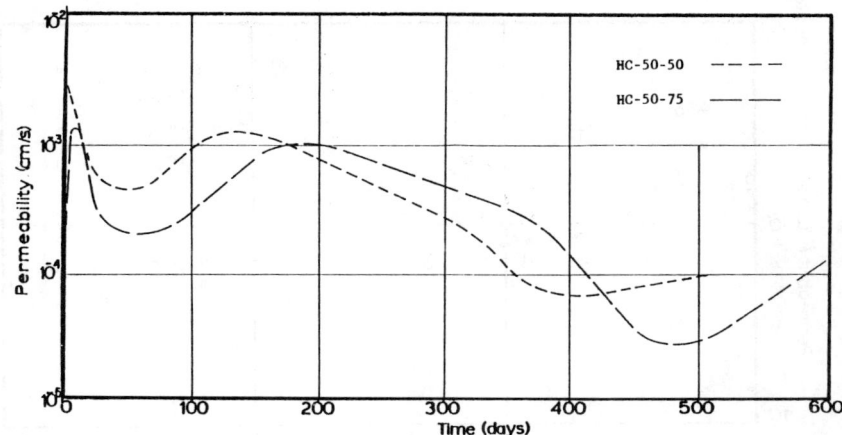

Figure 12. Effect of Non-neutralized Soda on Long-term Permeability

grained and fine-grained specimens were thoroughly scraped of all foreign material at the inlet end and fresh ungrouted Ottawa 20-30 sand filters were placed at both ends. Upon reapplication of the gradient, the "cleaned" specimens showed permeabilities similar to their peak permeabilities prior to cleaning; these permeabilities began to decline with time, as had been observed in the initial tests.

As a consequence of these observations, all of the galvanized iron pipes and fittings in the test manifold were replaced with non-corrosive brass fittings and nylon tubing and a filter was fitted to the main supply line prior to the initiation of Phase 2. To evaluate the effectiveness of these improvements, an experiment was conducted with specimen HF-60-50 from Phase 1. The end filters on the specimen were replaced with clean sand and all extraneous material was scraped from the inlet face of the grouted specimen; then, the specimen was returned to the manifold for continued testing. This process was repeated every thirty days. From Figure 13 it can be seen that most of the permeability decline was due to the extraneous material that accumulated in the end filters, as evidenced by the close correspondence of the data for each cycle. The scatter at the beginning of the cycle following the first cleaning is believed to be due to the fact that the specimen was stored in a sealed condition for several months prior to this cycle, and this could have resulted in an unsaturated flow condition or slight dehydration of the grout. These data also indicate that these improvements to the test apparatus were unsuccessful in completely eliminating foreign material from the supply water. However, the lack of heavy staining in the coarse-grained samples in Phase 2, as was observed in Phase 1, indicates that some success was achieved.

Observations from Dismantling Specimens

Upon termination of the continuous flow tests, each specimen was dismantled and carefully examined. Of primary interest in these exam

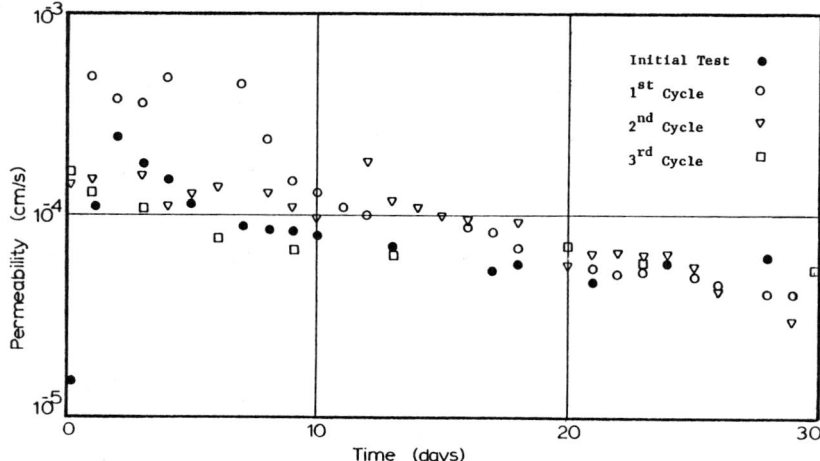

Figure 13. Influence of Test System Deficiencies on Permeability Observations

inations was the amount and distribution of any remaining grout and the occurrence of any anomalies, such as concentrations of foreign materials. These examinations involved both visual and microscopic observations and chemical analyses, and the extent of the examination procedures varied for each specimen.

AM-9, AC-400, and Sodium Aluminate Grouts -- When all specimens injected with AM-9, AC-400, or sodium aluminate grouts were dismantled, the grout appeared to be intact. There were no signs of erosion or other forms of grout loss and a uniform distribution of grout was observed throughout all specimens. Extraneous orange residue appeared in one of these specimens (gravel injected with sodium aluminate) and was confined to the inlet filter with concentrations appearing mainly near the walls of the permeameter. Since the grout appeared to be intact and uniformly distributed throughout this specimen and since it exhibited the highest permeability of all sodium aluminate specimens, it is reasonable to infer that much of measured flow probably occurred along the boundary.

Silicate Grouts -- Observations in dismantling specimens injected with silicate grouts (other than sodium aluminate) can be divided into two groups, based on the duration of exposure to flowing water. A series of five coarse-grained specimens treated with Hardener 600 and cured for up to six hours all exhibited Pattern 1 behavior, with termination and dismantling occurring immediately after the permeability jump. The specimens were allowed to air-dry at a temperature of 30° C for several days, after which all ungrouted sand could be easily removed with a probe. The typical cross-sections shown in Figure 14 indicate that the amount of grout remaining can be related to the amount of time the gel was allowed to cure prior to testing. The grout in specimens cured for ten minutes or less appeared to be completely elutriated, whereas the specimen cured for six hours had a uniform distribution of grout. Specimens with intermediate curing times showed elutriated zones that decreased in size with increasing grout age, suggesting that the adhesion between the grout and soil may increase with time. The configurations of the elutriated zones in these latter specimens also suggest that the elutriation process is initiated along the boundary. Similar configurations were observed in Geloc-4 specimens allowed to cure for short times prior to testing. Hardener 600 specimens cured for periods longer than six hours followed behavior Patttern 2 and were tested for extended periods.

For all silicate grouted specimens tested for extended periods, the material in the permeameter was found, upon dismantling, to have approximately the consistency of wet sand. No grout could be observed visually or microscopically, and a comparison of dry weight reductions due to washing the material with strong solutions of sodium hydroxide to dissolve any

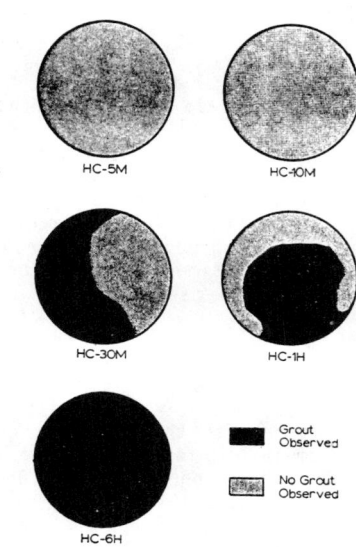

Figure 14. Regions of Elutriated Silicate Grout in Short-term Tests

remaining grout proved inconclusive. Drying undisturbed specimens at low temperatures (approximately 30°C) removed the pore water without causing excessive desiccation of the grout structure and prompted two observations. First, all specimens appeared to be lightly cemented throughout, indicating a thorough distribution of the remaining grout; since the grout could not be observed prior to drying and because similar cementation did not occur in disturbed material taken from the permeameter prior to drying, the remaining grout was probably concentrated at the contact points between the sand grains. And second, the concentration of the remaining grout, as indicated by the degree of cementation observed in the sand, decreased along the length of the specimen, with the minimum concentration at the outlet end. This implies that the erosion process is initiated at the outlet end of the specimen and proceeds upstream.

In several medium-grained and fine-grained specimens from Phase 1, a very hard layer, about one-half inch thick, was observed at the inlet face of the grouted sand. However, because of the presence of foreign materials, it is not clear if this hard layer represents the remaining non-eroded grout in the specimen or artificial cementation due to reactions with the foreign material. For example, it is very possible that this layer is the result of a low concentration of calcium in the tap water reacting with the ionized silicate to form a calcium silicate (Graf, 1984); this could also explain the cementation observed in the elutriated specimens.

CONCLUSIONS

Based on the results of this investigation, coupled with other data reported in the literature, the following conclusions can be drawn:

1. The polyacrylic grouts AM-9 and AC-400 show no signs of deterioration or erosion due to the application of a gradient of 100. The permeability of specimens injected with these grouts remained constant throughout the tests and appears to be independent of the curing time allowed prior to testing or the average void size in the soil matrix.

2. Specimens injected with the sodium silicate grouts underwent large variations in permeability during the early stages of testing, but, once the permeability stabilized, it remained relatively constant for the remainder of the test. The value at which the permeability stabilized appears to be dependent on the permeability of the ungrouted soil and, to a lesser extent, on the chemical characteristics of the grout. In general, it appears that, at gradients between 50 and 100, the maximum long-term reduction in the permeability of a soil due to the injection of these grouts is one to two orders of magnitude.

3. The amount and rate of grout elutriation appear to be dependent on the strength of the gel and the amount of syneresis experienced in the grout. The polyacrylic and sodium aluminate grouts, which achieve their maximum strength very shortly after gelation and experience little or no syneresis, showed no sign of grout elutriation when cured for periods of at least ten and thirty minutes, respectively. The silicate grouts require several days to achieve their maximum strengths and exhibit reductions of as much as 25% of their original volume due to syneresis. Specimens injected with silicate grouts and cured for less than one day experienced rapid, and usually complete, elutriation due to lack of strength. In older specimens the elutriation was a gradual process and the rate at which the permeability increased was apparently accelerated by increases in the degree of syneresis.

4. For the silicate grouted specimens in which most of the grout was eroded, that grout which remained appeared to be concentrated at the contact points between the soil grains.

5. Extreme caution is recommended when considering the silicate grouts tested in this program for use in situations where they will be subjected to high gradients.

When evaluating the results of this study, several important conditions inherent to this test program must be fully recognized and appreciated. First, the gradients employed were very high -- well above those normally encountered in most engineering applications; although there is limited evidence to suggest that the same general phenomenon occurs over a prolonged time frame at gradients on the order of 10 or less, there may exist some maximum gradient below which these grouts will not deteriorate with time. Second, the permeant (and mixing water in most cases) was tap water from the municipal water supply system of Evanston, Illinois; this water is certainly very different in composition than many of the groundwaters found throughout the United States and the world, and the influence of hardness, pH, or chemical constituents may be significant. For example, water with an acidic pH should serve to keep the silicic acid gelled, whereas water with an alkaline pH, which has the potential to dissolve the silicic acid gel, will frequently be rich in calcium carbonate or calcium sulphate and the calcium ions will react with the silicate ions to form a strong calcium silicate; in such a situation the permeability may increase somewhat, but the strength of the reaction product should be high. Third, of the dozens of "silicate grouts" that are available, only five were examined in this study; while many of the commonly used silicate grouts are of the silicic acid gel type utilized in these tests, several others (such as the Joosten process, which employs sodium silicate and calcium chloride in a flash reaction) rely on different reactions and exhibit different physical and chemical characteristics. And fourth, in certain field situations that were approximated in this laboratory study (such as the imposition of a gradient on a newly installed grout cut-off in coarse sand or gravel), a grout with a very short gel time (perhaps 30 seconds or one minute), or even a cement grout, would be used initially and then followed by a second phase of grouting; since the strength gain curves for short-gel-time grouts may be different from those for the longer-gel-time grouts used in these tests, the observations reported herein must be viewed in this light. Notwithstanding the findings of this study, the cautions that are suggested, and the seriousness of potential problems that may result, it is well known that many field applications of various silicate grouts have withstood the test of time for decades with no evidence of deterioration under conditions of quasi-static groundwater or seepage under very low gradients. The challenge is to learn within a reasonable degree of certainty when and why a particular grout will or will not perform satisfactorily.

ACKNOWLEDGEMENTS

Grateful appreciation is extended to Wallace H. Baker of GKN-Hayward Baker Company, Edward D. Graf of Pressure Grout Company, and William J. Clark of Geochemical Corporation for their stimulating discussions and constructive criticisms during the progress of this study and preparation of this paper.

REFERENCES

Baker, W. H. (1982), Personal Communication.

Bentayf, M. A. (1981), Effective Stress-Strain-Strength of a Silicate Grouted Sand, Ph.D. Dissertation, Department of Civil Engineering, Northwestern University, Evanston, Illinois, pp. 28-48.

Cambefort, H. (1964), Injection des Sols, Volume I: Principles et Methodes, Editions Eyrolles, Paris.

Cambefort, H., and Caron, C. (1957), "The Leaching of Sodium Silicate Gels," Proceedings of the Fourth International Conference on Soil Mechanics and Foundation Engineering, Volume I, London, pp. 13-17.

Celtite, Incorporated (1980), Celtite Chemical Grouting Products for Foundation Engineering, Cleveland, Ohio, pp. 1-7.

Christopher, B. R. (1979), Evaluation of Specimen Preparation and Testing Procedures for Chemically Stabilized Granular Materials, M. S. Thesis, Department of Civil Engineering, Northwestern University, Evanston, Illinois.

Clark, W. J. (1982), "Performance Characteristics of an Acrylate Polymer Grout," Proceedings of the Conference on Grouting in Geotechnical Engineering, American Society of Civil Engineers, New Orleans, Louisiana, pp. 418-432.

Davidson, R. R., and Perez, J. Y. (1982), "Properties of Chemically Grouted Sand at Lock and Dam Number 26," Proceedings of the Conference on Grouting in Geotechnical Engineering, American Society of Civil Engineers, New Orleans, Louisiana, pp. 433-450.

Einstein, H. H., and Schnitter, G. (1970), "Selection of Chemical Grout for Mattmark Dam," Journal of the Soil Mechanics and Foundations Division, American Society of Civil Engineers, Volume 96, Number SM6, pp. 2007-2023.

Graf, E. D. (1984), Personal Communication.

Hurley, C. C., and Thornburn, T. H. (1972), "Sodium Silicate Stabilization of Soils: A Review of the Literature," Highway Research Board, Highway Research Record Number 381, pp. 46-79.

Karol, R. H. (1957), "Development of a New Chemical Grout," Proceedings of the American Society for Testing and Materials, Volume 57, pp. 1219-1232.

Merrill, R. C., and Spencer, R. W. (1950), "Gelation of Sodium Silicate," Journal of Physical and Colloid Chemistry, American Chemical Society, Volume 54, Number 6, pp. 806-812.

Plank, C. J., and Drake, L. C. (1947), "Differences Between Silica and Silica-Alumina Gels - I. Factors Affecting the Porous Structure of These Gels," Journal of Colloid Science, Volume 2, 4, pp. 399-412.

Plank, C. J. (1947), "Difference Between Silica and Silica-Alumina Gels - II. Proposed Mechanisms for Gelation and Syneresis of These Gels," Journal of Colloid Science, Volume 2, 4, pp. 413-427.

Rhone-Poulenc, Inc. (1975), Hardener 600 Series, Paris, pp. 1-60.

Tallard, G. R., and Caron, C. (1977), Chemical Grouts for Soils - Volume II, Report to Federal Highway Administration, U. S. Department of Transportation, Washington, D. C.

GROUT PENETRABILITY
R.H.Karol, P.E., Member ASCE*

ABSTRACT

The ability of a grout to penetrate a formation is always a selection requirement for projects in which fracturing is undesirable. The major grout properties affecting penetrability are grain size (for suspended-solids grouts) and viscosity (for true solution grouts). This paper summarizes the comparative positions of common commercial grouts with respect to each other, and suggests a modification in the manner in which comparative data is presented.

Grouts must be able to permeate formations requiring treatment at placement rates which make those grouts cost effective. Penetrability is thus a primary factor in the selection of a grout for a specific application. For grouts which consist of solid particles suspended in a liquid (such as cement and clay), the relationship between the particle size of the grout and the void size of the formation determines the penetrability. For grouts which are solutions containing no solid particles (such as acrylates and phenols), the relationship between viscosity and void size determines the penetrability. (Silicate grouts are actually colloidal solutions, but act like solution grouts).

With few exceptions, the voids in both soils and fractured rock formations are distributed through a range of sizes, which may be narrow for uniform granular deposits and wide for heterogeneous granular deposits and fractured rock formations. Thus, the void size used as an index of penetrability of a formation is neither its maximum nor its minimum, but some value in between those limits. Since each soil deposit is unique, all attempts to specify void sizes (such as the D_{50} notation) are in reality generalizations, despite their appearance of specificity.

*Professor Emeritus, Rutgers University, Department of Civil Engineering, New Brunswick, NJ 08901.

Groutability data have appeared in several publications:

Caron [1] summarizes data from a Russian paper:

1) Coarse sand (greater than 0.8 mm) can be injected by suspensions containing particles up to .50 microns.

2) Medium sand (from 0.1 to 0.8 mm) can be injected by colloidal solutions.

3) Fine sands and silts can be injected by newtonian solutions of low viscosity.

Scott [2] states that a grout containing even a small proportion of coarse particles can form a filter cake in the soil face near the injection source (thus effectively halting the grouting process). He further states that cement suspensions with particles as large as 100 microns will form filter cakes in soils with permeabilities as high as 10^{-2} cm/sec, and suggests a method for relating maximum permissable grout size to pore size. (These data relate to permeation, not fracturing. All the discussion which follows is also related to permeation).

Mitchell [3] defines groutability ratio for particulate grouts pumped into soils as $\dfrac{(D_{15}) \text{ soil}}{(D_{85}) \text{ grout}}$ and states that successful grouting requires this ratio to be greater than 25. Similarly for rock grouting, the groutability ratio is defined as $\dfrac{D \text{ (fissure)}}{(D \text{ max}) \text{ grout}}$ and successful grouting requires ratios greater than 3. This is often stated as a criterion for avoiding "blinding" (the blockage of open passages), and it is easy to visualize three cement particles attempting to enter a void simultaneously, and blocking it.

Groutability ratios are applicable only to particulate grouts permeating granular deposits, since specific grain size data is needed for both the grout and the formation to compute the ratio. Using such data, it is possible to establish mathematically the penetrability limits for cement and most clay grouts. (for bentonite grouts, a viscosity factor must also be considered). For ordinary Portland cement, coarse sand is generally recognized as the lower limit of penetrability.

Chemical grouts which are true solutions have no grain size data to apply to a groutability ratio formula. In theory, any solution can be pumped into any porous formation, providing no limits are placed on pumping time or pressure. In practice, however, there are both safety and economic limits that must apply. These may be vastly different for the same soils under different conditions. Depth of treated formation, which affects allowable pressure, and cost of labor which affects job costs, are factors which affect economic feasibility, as opposed to theoretical feasibility. Thus, determination of the penetrability limits of chemical grouts is far less precise than estimates for particulate grouts. Nonetheless, groutability charts have appeared in manufacturer's literature and technical publications as early as thirty years ago. Although the penetrability limits shown in publications may in some cases be open to question, the relative positions of these limits for different grouts are almost always shown in proper perspective. This is because the theoretical expressions for flow of a liquid through a pipe or tube (such as Poiseuilles Law $V = \frac{\pi p\ r^4 t}{8\ L \mu}$, all indicate that flow rate varies inversely with viscosity.[4] For any specific limiting pressure, higher viscosities result in longer pumping times, and higher costs. Thus economic feasibility is inversely related to viscosity (i.e. the lower the viscosity, the better the economic feasibility as related to labor costs.) Figure 1 shows a recent compilation made by the author of various methods of improving soil properties, related to the soil grain szie and permeability characteristics in which those methods are applicable. (In some cases, upper limits as well as lower limits are shown.)

A study of the literature shows many case histories in which penetration well below the lower limits shown in Figure 1 have been attained. These improvements in penetration have variously been ascribed to the use of "slippery" grouts, the pre-injection of lubricating liquids and the use of surfactants or other additives. It is the author's opinion that surfactants can at best cause only a very mild improvement in grout penetrability, and that a viscosity reduction produces far more significant changes. The anomalies reported in the literature are most probably due to injection into very loose formation and/or the use of grouting pressures high enough to induce fracturing.

Among the most recent data published by product manufacturers is the chart shown in Figure 2. This chart is specifically intended to show the penetrability of a new product (MC-500) compared with chemical grouts. Comparing the D_{85} of MC-500 and ordinary Portland cement yields a ratio of $\frac{2.5}{8}$ = about 0.31. Comparing the D_{15} values yields a ratio of $\frac{5.8}{45}$ = 0.13. These ratios average

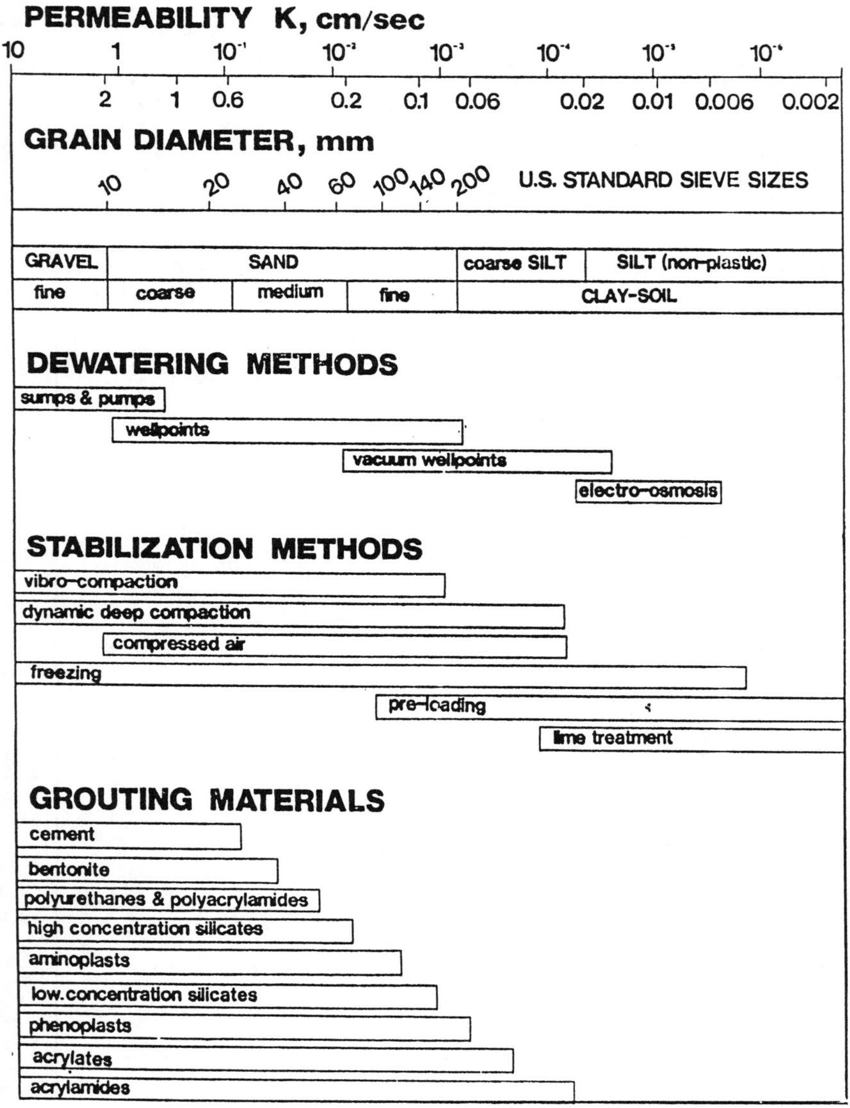

PERMEABILITY K, cm/sec

10	1	10^{-1}	10^{-2}	10^{-3}	10^{-4}	10^{-5}	10^{-6}
	2 1 0.6	0.2	0.1 0.06	0.02	0.01	0.006	0.002

GRAIN DIAMETER, mm

10   20   40   60   100 140 200   U.S. STANDARD SIEVE SIZES

GRAVEL		SAND			coarse SILT	SILT (non-plastic)
fine	coarse	medium	fine			CLAY-SOIL

DEWATERING METHODS

- sumps & pumps
- wellpoints
- vacuum wellpoints
- electro-osmosis

STABILIZATION METHODS

- vibro-compaction
- dynamic deep compaction
- compressed air
- freezing
- pre-loading
- lime treatment

GROUTING MATERIALS

- cement
- bentonite
- polyurethanes & polyacrylamides
- high concentration silicates
- aminoplasts
- low concentration silicates
- phenoplasts
- acrylates
- acrylamides

Figure 1

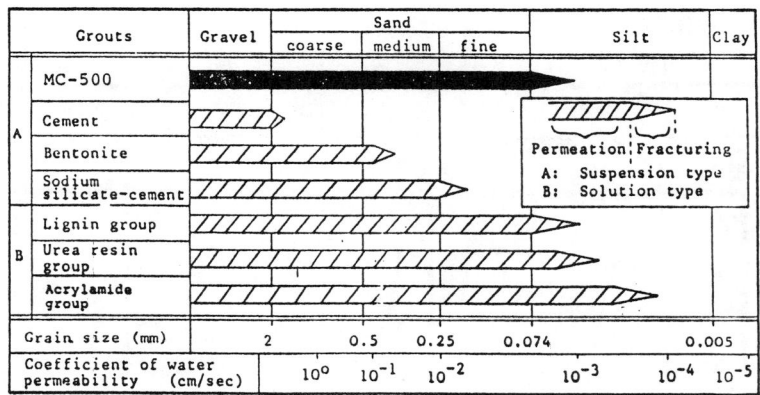

Figure 2.

0.22. If we accept the cement limit shown in Figure 2 as above coarse sand (20. mm), then the lower limit for MC 500 should be (using Mitchells ratio) 2 x 0.22 = 0.44 mm, in the medium sand range. If we accept the cement limit shown in Figure 1 as above medium sand (0.6 mm), then the lower limit for MC 500 should be in the middle of the fine sand. In either case, it appears that the average penetrability difference between ordinary Portland cement and MC-500 has been somewhat exagerated. The chart in Figure 2 may be correct or even conservative in depicting maximum differences.

The conflicting claims and data regarding grout penetrability can be confusing to purchasers of grouting services, and could become an unnecessary burden to the grouting profession. The confusion and conflicts could be at least partially resolved by a different method of data presentation. The two major factors which influence penetration into a specific granular soil type (such as medium sand) are formation density and percentage of fines (-200 sieve size). The latter factor can be accounted for by presenting the limiting penetration as a sloping line on a grain size accumulation chart. The plotted data should represent medium densities, and a note on the chart can discuss density differences.

A suggested starting point for chart development is shown in Figure 3. The data is considered tentative, and should be revised by field experience. There is obviously a question as to the proper slope and parallelism of the

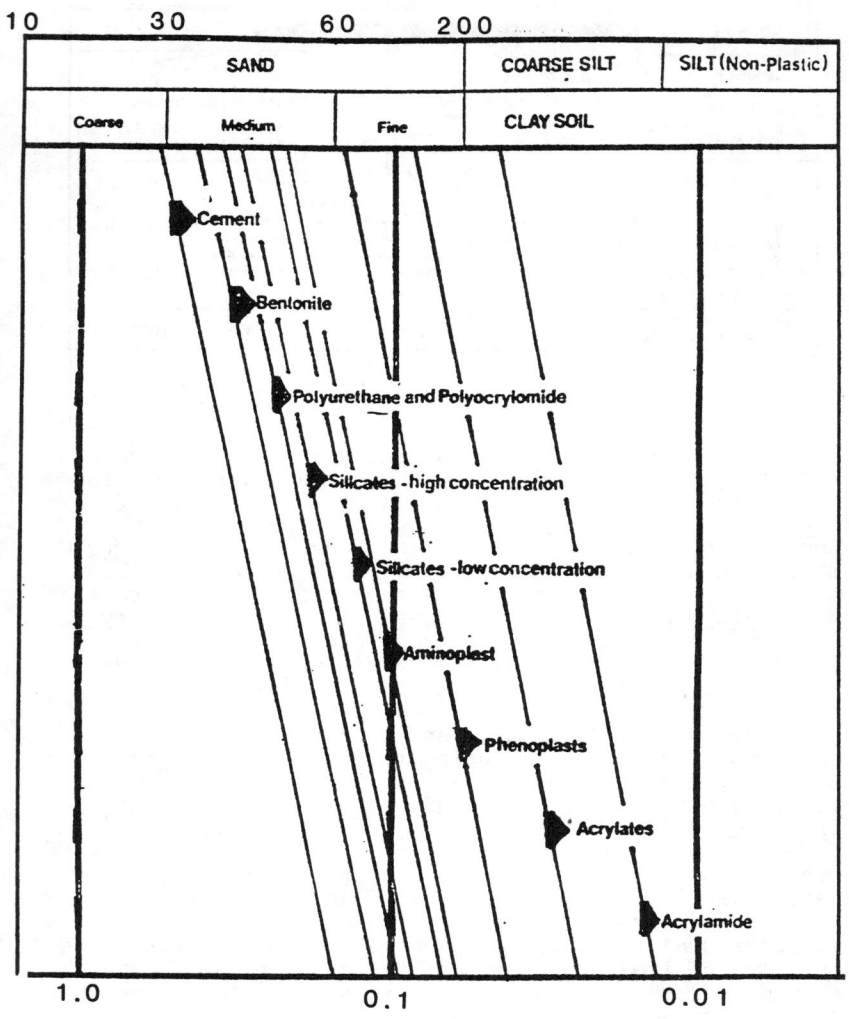

Figure 3

of the lines as shown, and even as to the shape of the enclosed zones (should the lines be curved, rather than straight). Further, there is a question as to whether percents passing above 50% have any significance.

The author hopes that constructive comments from readers will lead to more accurate representation of penetrability.

REFERENCES

1. Caron, C. "Grouts and Drilling Muds in Engineering Practice". pp. 98-103, Butterworths, London, 1963.

2. Scott, R.A. "Grouts and Drilling Muds in Engineering Practice", pp. 4-8, Butterworths, London, 1963.

3. Mitchell, J.K. "In-Place Treatment of Foundation Soils" pp. 73-109, Journal of the Soil Mechanics and Foundations Division, ASCE New York, January 1970.

4. Karol, R.H., "Chemical Grouting, pp 33, Marcel-Dekker, New York, 1983.

CEMENT GROUTING: WATER MINIMISING PRACTICES
A. Clive Houlsby, F. ASCE *

ABSTRACT

Water gives cement grout its mobility but can become a hindrance once its transporting role is over.

The water needed for mobility exceeds hydration requirements substantially in most applications. If this excess water is allowed to collect and remain in pockets, it can impair the production of 100% grout filling and may permit long-term deterioration.

Construction expedients to minimise the problem have been developed over years of experience by the author and others. Some aspects of these do not appear to be widely known (bleeding methods) and others (w:c ratios), although discussed, have received little presentation in the context of the whole gamut of cement grouting. This paper therefore reviews the issue. Tests are described, and case histories quoted.

INTRODUCTION

The chemical action of cement hydration requires a water:cement (w:c) ratio of less than 0.3:1 by weight or 0.45:1 by volume. Water used in excess of this is only for the purpose of making the grout workable and transporting and placing it.

Ideally, once grout has been injected into spaces it should not shrink and leave less volume of solid filling than that of the initial liquid grout. To achieve this involves retention of excess water in the set product. However, in most types of cement grouting the amount of water used exceeds the capacity for such retention and therefore may need to be disposed of. The ideal situation is rarely achieved but the retention of some excess water in hardened grout is normal. An axiom of grouting is to use the minimum amount of water possible consistent with the required mobility.

> * Engineer Specialising in Grouting, and Senior Surveillance Engineer, Water Resources Commission, North Sydney, N.S.W., Australia

BEHAVIOUR OF EXCESS WATER

The basic issue is that of bleeding. Once the movement of grout slows or the grout is motionless, cement particles endeavour to gravitate downwards while water and the thinner fractions of the grout tend to rise. The process can be accompanied by density currents in large bodies of grout. Given thin grouts and undisturbed conditions for long enough, cloudy or even clear water can collect above the fairly distinct surface of settled cement. Fig. 1 compares the amounts of bleedwater rising from various initial thicknesses of cement grout without additives or sand. Even a mix in common usage for foundation grouting, such as 2:1, by volume can develop 35% of bleedwater eventually, leaving only 65% of spaces filled with cement unless the bleedwater can be removed and replaced with more grout. For this mix the bulk of the bleeding takes an hour or so, in contrast to very thin mixes such as 12:1 where the first fifteen minutes sees a rapid separation of water and cement.

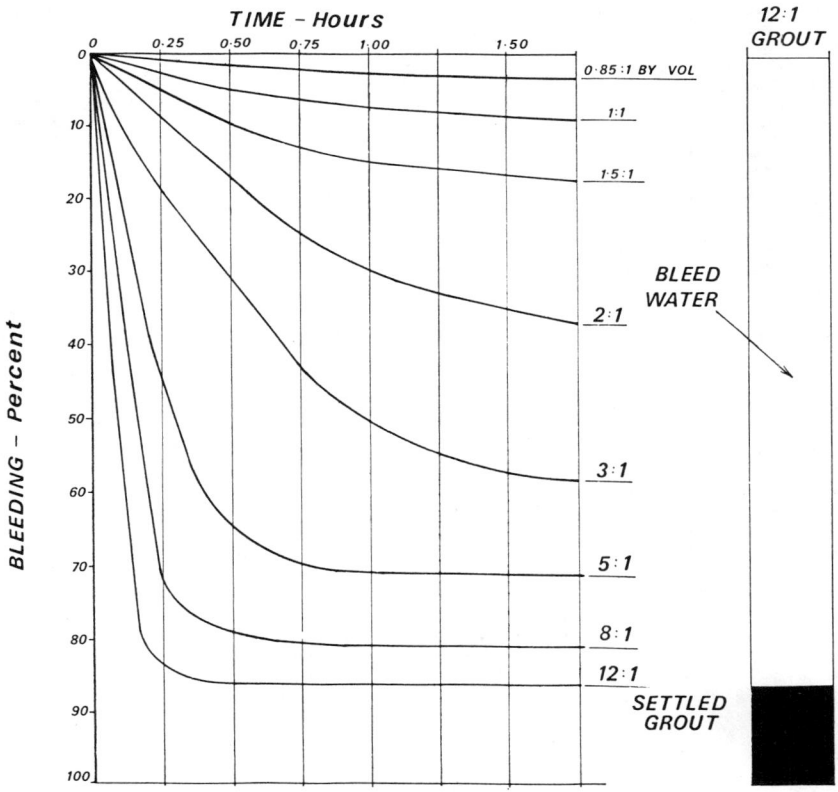

FIG. 1   Percent bleedwater rising from grout.

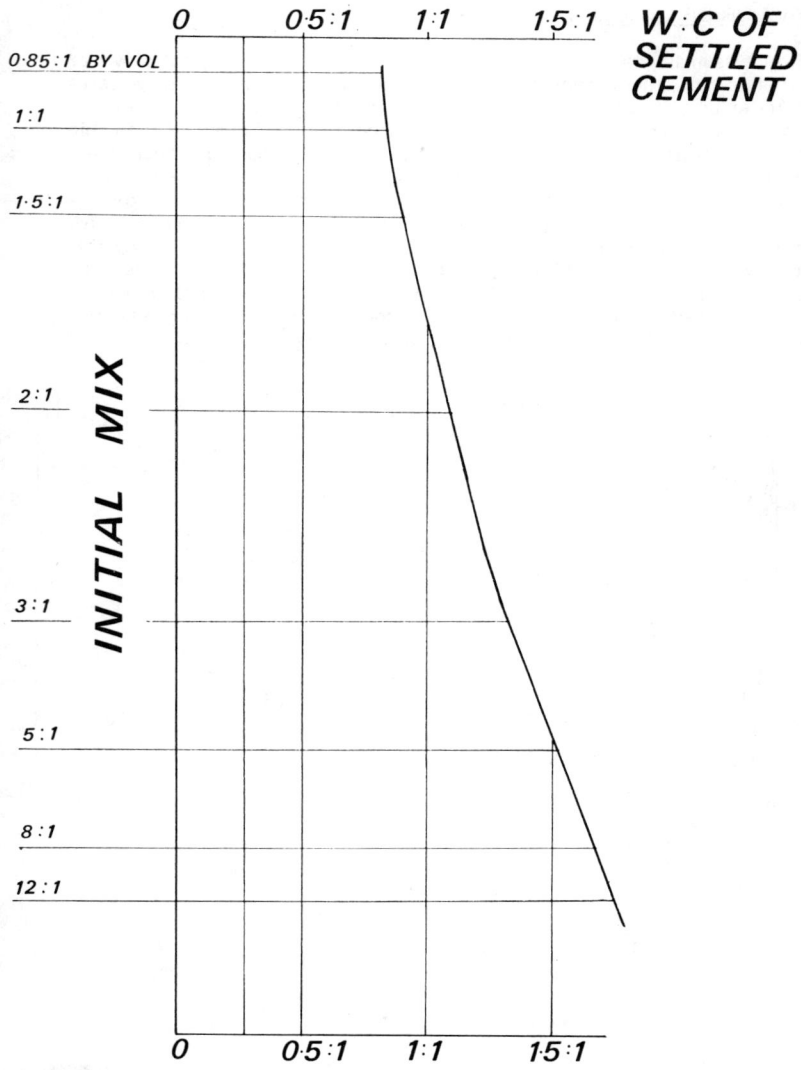

FIG. 2 Density of settled part of grout.

The final density of the settled cement varies, depending on the initial w:c ratio. Fig. 2 attempts a crude portrayal of this by indicating the approximate w:c ratio of the settled cement and comparing it with the initial w:c ratio. A 2:1 grout can be expected to settle approximately to 1:1 while 12:1 could settle to about 1.6:1 given adequate quiescent void space and time. This variation of density is a measure of the amount of excess water retained in the settled fraction. The physical appearance and strength of the set grouts at the thin end of the scale are poor and their durability is suspect as discussed later. Further, their setting time can be surprisingly long. A 5:1 grout can take 24 hours to reach initial set in the settled fraction and 12:1 grout can take a matter of days or even weeks. Thus these thin grouts could easily be washed out of position before setting, by water or grout from subsequent holes nearby in a foundation grouting operation. They can also be flushed away by flowing groundwater in some sites below the water table.

In order to minimise settlement, some grouting practicioners use a small quantity of bentonite and describe the resulting grout as "stable". Opinions vary as to the effectiveness of this and whether it produces a weaker grout than if no bentonite is used. The writer has found the use of bentonite not necessary, provided first class grout mixers are used and proper injection techniques applied. This paper therefore does not refer to grouts containing bentonite.

GROUTING IN A ROCK FOUNDATION

Cement grouting of open cracks and voids in rock is predominantly for purposes of controlling seepage or for strengthening. Durable work is needed usually. The proportion of open cracks required to be filled varies depending on the standard of grouting required for the project, (3) and (5). At times there is a need to thoroughly fill every crack, but not often. A commoner situation requires only larger cracks and voids to be filled, or possibly requires a network of fine cracks to be partially filled. In these part-filled situations, bleedwater can escape to the non-grouted cracks frequently. The grout filling in those parts which are grouted must, however, be thorough to avoid being leached away by seepage in subsequent years.

Some cracks are easier to grout than others. The easiest are near-vertical; the hardest are those which are near-horizontal and have no cracks leading upwards from them. The near-vertical ones are easiest because the cement tends to settle uniformly in them and bleedwater can rise up and escape to the surface or other convenient egress. The wider the crack the better this happens: thin cracks can have hang-ups due to surface tension, wall roughness, ridging, separation in the grout, etc. Cracks of a horizontal orientation can be hard to grout thoroughly if the bleedwater is unable to escape through cracks leading off upwards. Fig. 3 illustrates easy and difficult cracks.

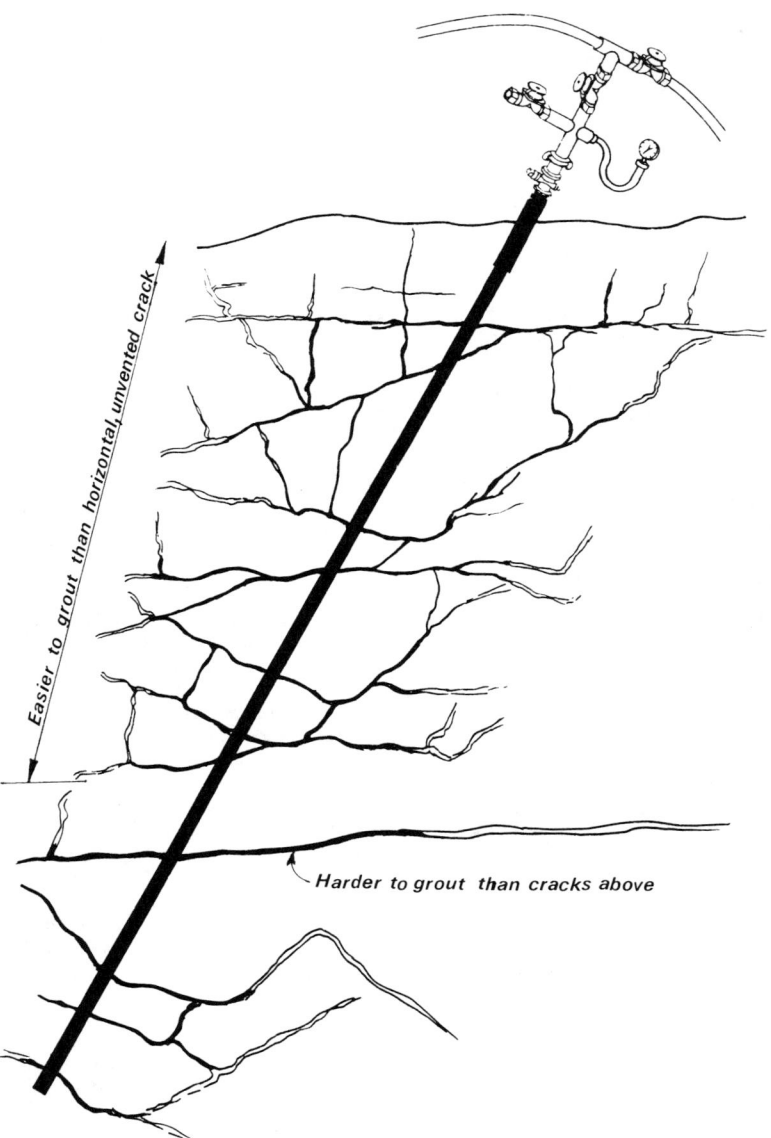

FIG. 3. Cracks of differing grouting ease.

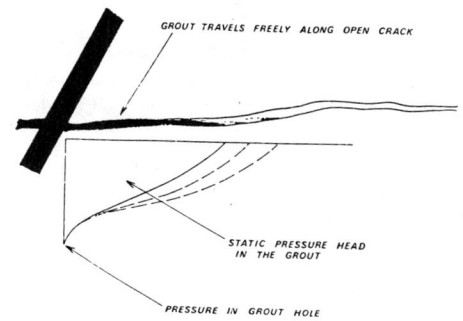

FIG. 4. Mode of grout travel in horizontal crack

Fig. 4 is an enlargement of the mode of grout travel along one of the difficult cracks. At the start of grouting (step (1)) the grout travels from the grout hole freely along the open crack under the action of pressure from the grout hole. The pressure falls off as the distance from the hole increases until at the tip of penetration it is relatively slight. As the grout continues its penetration along the crack (step (2)), the pressure profile follows in much the same way until the limit of penetration is reached for the particular pressure and w:c ratio used (step (3)). Up to this stage grout further back in the crack has been in motion and additional grout has been entering the crack from the hole. Hence there has been little opportunity for the bleeding to start. However, the cessation of movement now permits separation of water from cement. Because there is little room for this to occur in the upward direction, it takes place in a lateral fashion leaving ridges of grout separated by meanders of bleedwater. There is a visual resemblance to delta formation at the mouths of large rivers. This separation commences at the fringe of penetration where there is a surfeit of water in any case from groundwater and maybe residual water from the water testing as well as the bleed from the grout. The separation then progresses back towards the hole. The grout has started to stiffen and might no longer be able to transmit pressure from the hole. Hence the bleedwater is trapped amongst the grout and an imperfect filling results.

If there are frequent cracks venting upwards, the bleedwater tends to escape through them and this permits fresh grout to enter from the hole. However, if there are not, as illustrated in Fig. 4, the bleedwater can be trapped amongst the grout in an imperfect filling once it becomes too stiff to further transmit pressures from the hole.

A fuller description of the rheological behaviour of grout during injection is given in (5). This description comes from a combination of the tests of the U.S. Corps of Engineers (8), similar tests carried out by the Water Conservation and Irrigation Commission, New South Wales, Australia (described below) in formation from foundations excavated after grouting, and experience.

The foregoing relates to grouting carried out at moderate pressures so as not to deliberately spring cracks open by grout pressures. The alternative method of high pressure grouting, known as displacement grouting, claims to squeeze the bleedwater out. There are differing schools of thought about the wisdom or efficacy of this method, as outlined in the keynote paper of the New Orleans Grouting Conference (5). Subsequent discussion on the issue during the Conference proved to be meagre and use of the displacement method found little support amongst those present.

The mode of pressure distribution in a horizontal crack depends, amongst other things, on the crack width. A fuller treatment of the issue was presented at (2).

CEMENT GROUTING

The degree to which bleedwater is trapped amongst hard grout depends, of course, on the amount of water initially in the grout. The practice recommended in (9) of using 2:1 for most foundations with 3:1 as the thinnest mix in fine cracks recognises this.

BLEEDING AT THE GROUT HOLE

During the course of a normal grout application, water and thin grout accumulate in the upper part of the grout hole. Good practice is to release this through a bleed valve mounted amongst the control fittings on top of the hole. Fig. 5 shows a suitable arrangement.

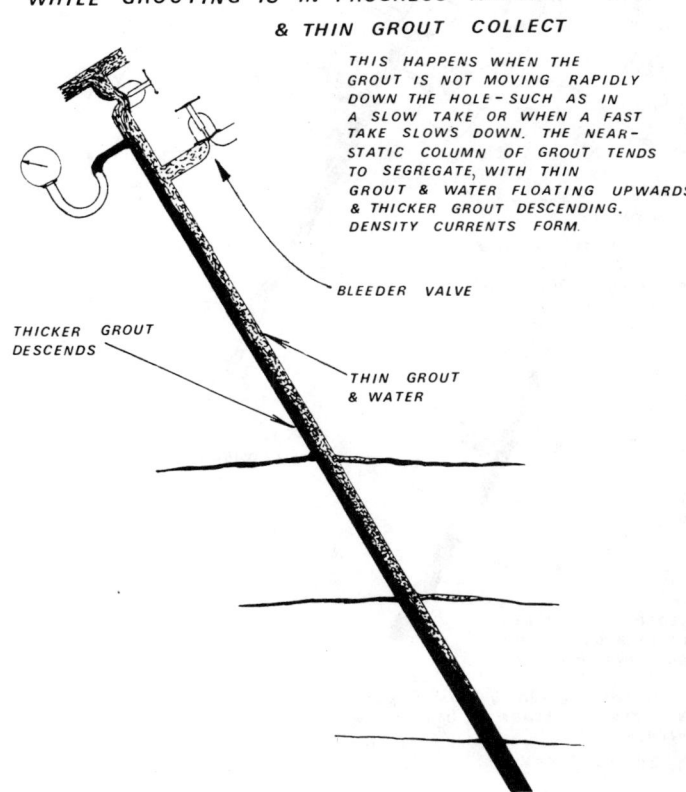

FIG. 5. Accumulation of bleed water in grout hole.

By bleeding off these sub-standard fluids, their place can be taken by good quality grout. The steps when bleeding are shown in Fig. 6 which comes from (9).

ISSUES IN DAM GROUTING

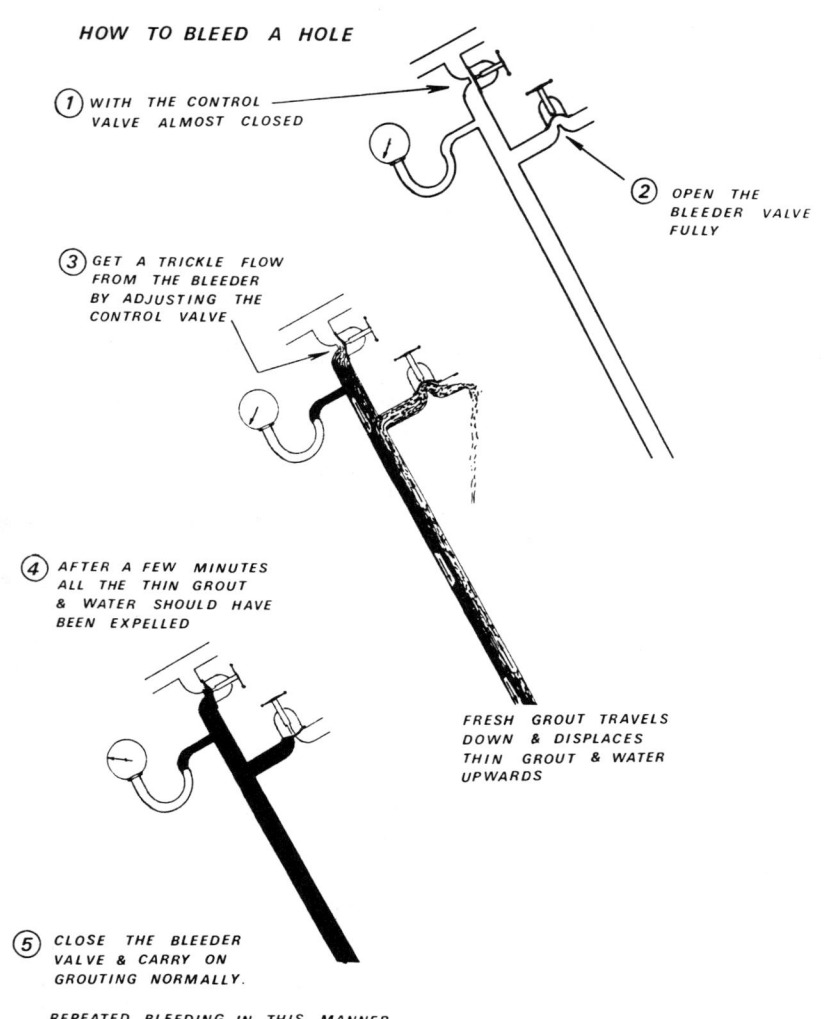

FIG. 6. Removing bleed water from a grout hole.

Bleeding is essential when the rate of take of grout in the hole slows down to the extent that conditions are sufficiently quiet for the water and thin grout to rise upwards while the thicker fraction of the grout descends. This condition commences in most holes towards the end of the grout application when the take becomes slow as refusal is approached. It can commence earlier if the take is insufficiently fast to keep grout in the hole in turbulent motion or if the hole is of large diameter. The typical percussion drilled grout hole of 2-1/2" diameter is really too big to give turbulent motion with the normal range of takes and hence bleeding from the commencement of the grout application has been found advisable. A mixed situation can develop if a large crack part way down the hole takes most of the grout supplied. The section of hole below it then tends to become filled with settled grout and the bleedwater from this can get carried off into the large crack. Such (beneficial) loss into the cracking system of bleedwater from grout holes is believed to be frequent but in most foundations is too random and unpredictable to be relied on. Hence routine bleeding is prudent no matter what benefit in this respect can be obtained from the foundation itself.

The frequency of bleeding should be at about fifteen minute intervals and should continue even while holding pressure after refusal. The amount drawn off at each bleed must be sufficient to clear fresh grout out of the section of supply pipe between the bleeder valves and the hole, plus the volume of water and thin grout, together with a small quantity of good grout from the hole to confirm that the bleed products have indeed been removed. The total volume of this drawn-off fluid can at times be enough to influence the measurements of grout taken into the hole and in such cases the volume of this fluid should be measured by collection in a bucket or similar. In contract work it is advisable to pay for cement lost by bleeding so as to encourage the contractor to bleed adequately.

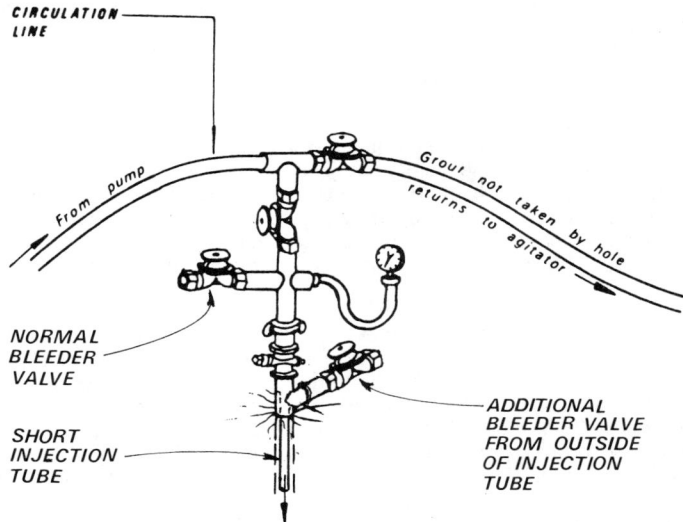

FIG. 7. Improved bleeding-off arrangement.

When packers are used, bleeding is difficult to carry out properly. This is because the relatively small diameter tube which carries the grout through the packer is too small to permit fresh grout to pass downwards while at the same time allowing bleed products to float upwards. There is too much turbulence in the tube during most of the grout application. It is only when refusal is being approached that sufficiently quiet conditions develop for release of the water and thin grout. They remain trapped below the packer until then unless cracks are fortuitously located appropriately to accept them. This is one of the reasons for the writer's aversion to the use of packers in general.

An improvement on the bleeding fittings shown in Figures 5 and 6 has been developed at Cressbrook Dam, Queensland, Australia, and was used with success for the grouting of that dam. It is shown at Figure 7. Separation between the incoming fresh grout and the rising bleed products is provided and a separate bleed-off valve connects directly with the bleed collection area. A normal bleed valve is also provided for the subsidiary purpose of checking for blocked fittings, checking the supply of grout and wash-out of gear.

Bleeding of grout holes is most effective with the thicker grouts. It is practical for grouts as thin as 3:1 but is probably of little benefit for grouts thinner than about 5:1. This is because these very thin grouts (5:1 and thinner) are, in their fresh condition, not much denser than the discarded bleed products from the thick grouts and also because of the difficulty of visually distinguishing between these grouts and their bleedwater in the spray emerging from the bleeder valve. Such identification is always necessary when deciding when to turn the bleeder valve off. A visual identification of 3:1 and thicker mixes is possible with experience but mixes thinner than this are difficult.

A simple routine method for checking whether bleeding has been carried out properly is to examine the grout hole a day or so after grouting. If the grouting has not left the hole substantially full of grout then the bleeding has probably not been adequately done.

The question can well be asked as to whether there is a real need to pay all this attention to the grout hole itself when the purpose of the grouting after all is to fill cracks and voids, not just merely the hole. In answering the question attention must be drawn to the role of the hole as part of the grout supply system. It is called upon to provide the designated w:c ratio to all cracks intercepted by it. If not bled, the situation could develop where upper cracks have only weak, watery grout available to them instead of the intended thicker mix.

Even with adequate bleeding, there is a tendency for grout to settle in the bottom of the hole and produce a density there which is greater than that of the mix supplied. If cracks in this region are too fine to accept this denser mix, then grouting may be impaired. This is one of the factors to be taken into account when deciding which thickness of mix to supply and when to thicken it during progress of the grouting. Such decisions devolve on experience largely along the basic lines given in (9). This matter of dense grout in the bottom of the hole is a function of the length of the hole and in stage grouting is one of the main factors determining the length of the stages. Unusually long stages can give excessively thick grout as the result of settlement.

The bleeding and settling behaviour of grouts is experienced to a strong degree when grouting vertical Contraction Joints in concrete dams. The substantial volume of bleedwater from mixes such as 2:1 and 1:1 has to be bled off through a vent system and replaced with grout and this may have to continue for several hours before all the thin grout and water are replaced.

During foundation grouting, unless bleeding off has been practiced and the hole filled satisfactorily with thick grout, there remain doubts about the efficacy of the grouting in cracks leading off from upper parts of it. If the grout in these upper parts was too weak to set firmly, or if it drained away by gravity after completion of the grouting, was it thick enough and durable enough to constitute effective grouting? The threat that it might not be can be particularly serious in the top few feet of a dam foundation where adequate grouting is more important than at depth. As indicated above, control of the potential problem can be obtained by use of bleeding off procedures and many dam sites have been grouted using them. However, on some sites where this has not been practised, there has been some disquiet about the implications of consistently unfilled holes. Instances at several dam sites in the early 1970s led to a series of tests of the behaviour of grout in the grout holes. These tests were described in a private communication to the writer from the late Charlie G. Flagg, lately of the U.S.B.R. and formerly of the Corps of Engineers. A description of the tests follows, with acknowledgement to Mr. Flagg..

BLEEDING TESTS IN SIMULATED GROUT HOLES

The tests simulated grout holes using 2-1/2" diameter plastic pipe, some of it transparent and the rest opaque. The grout was applied through a tremie tube and therefore the placement was not identical with the manner in which grout runs freely down the hole at the commencement of a normal grout application. However, once the hole has been filled, subsequent behaviour of the grout in the testing is regarded as representative of conditions in routine grout holes where unbled. The grout was mixed in a paddle mixer (low speed of rotation, low shear) and therefore showed greater bleeding and segregation effects than if high speed mixing with high shear had been used. The grout surface probably merged somewhat with water placed in the pipes beforehand.

Initial testing used a simulated vertical hole 32 ft. high, (Fig. 8) the lower 12 ft. being clear acrylic and the upper 20 ft. white opaque P.V.C. Draw-off valves were sited at various levels. 20 ft. of water placed in the pipe beforehand was displaced by 1:1 grout injected through the tremic tube.

Within thirty minutes of placement settlement and segregation could be seen in the transparent section. The lower 12 ft. developed a dark colour. Colour became lighter upwards in a variegated pattern. Fig. 8 shows the properties of the liquid grout from samples drawn off about forty-five minutes after placement. The specific gravity and strength of the grout in all but at the foot of the hole had bled to weaker values than when first placed. Fig. 9 shows where 5 inch voids of clear water developed after several hours and it also shows the gradual diminution of fluid pressure at the foot of the hole over the first four hours. Additional voids developed in the opaque section to within 3 ft. of the top. Obvious arching across the hole took place. Formation of this type of arched top on bleedwater pockets has been found elsewhere as reported in (4).

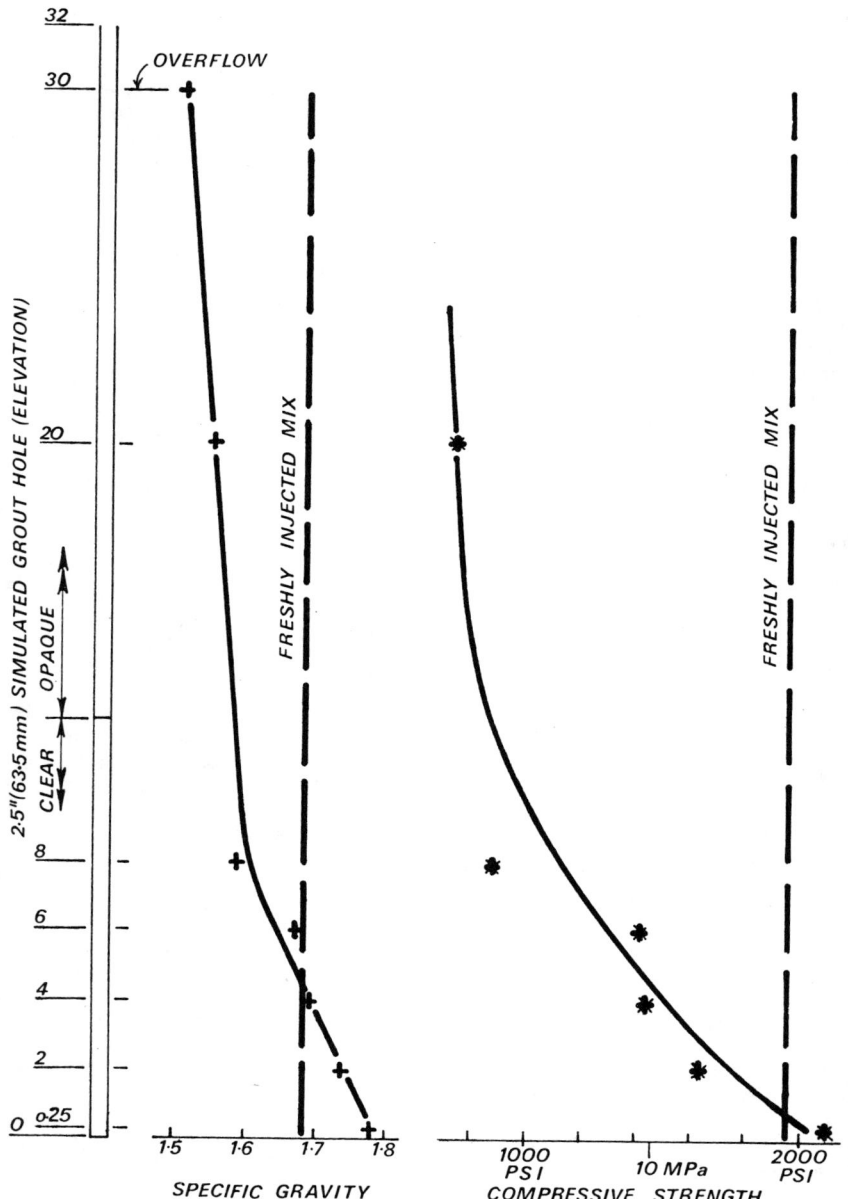

FIG. 8. Measurements in simulated grout hole (after C.G. Flagg).

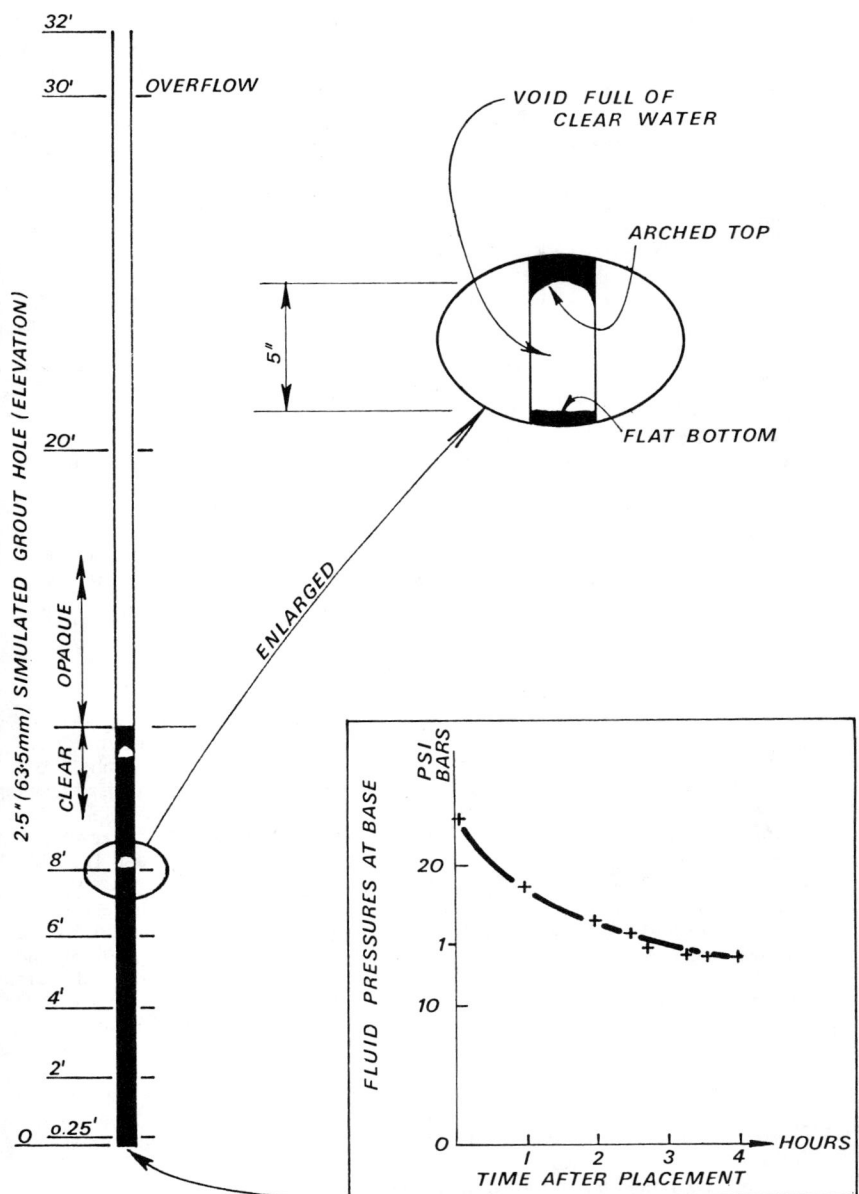

FIG. 9. Formation of voids in simulated grout hole (after C.G. Flagg).

These tests in a vertical tube prompted tests in inclined ones. Pipes were 30 ft. long at 30° off the vertical. 3:1 grout was used in one and 1:1 grout in the other. The grout was left undisturbed for fourteen days and the tubes were then cut longitudinally. The opened up samples showed that

- voids were not as concentrated or as large as in the vertical test

- numerous small voids or bubbles were spread along the upper side of the slope. These were more numerous in the 3:1 grout and were up to 0.75 inches diameter.

- the quality of the grout varied in cross-section. Channels on the upper side of the slope contained the less dense grout.

- variations in the colour of the grout indicated appreciable density current formation and segregation.

One of the recommendations made from this testing was that high speed mixing should be used. It was realised that the use of paddle mixing contributed to the poor results and was partially responsible for the noticeable segregation and, to a lesser extent, the amount of bleedwater accumulation. Grout should be uniform in cross-section when set, even down in the hole, and any mixer which does not routinely produce such uniform grout should not be permitted. Some commercially available brands of high speed mixers comprising a 2,000 R.P.M. (approx) rotor revolving in its own chamber and circulating through a separate vortex, are able to produce excellent grout.

EXAMPLE OF BLEED WATER TRAPPED IN GROUT HOLE

An illustration of the manner in which water can be left trapped in a grout hole if it is not thoroughly bled off is shown in a photograph by Cambefort in (1). Pockets of non-grout filled hole appear as voids in a core of hard grout taken from a hole. Distinct streaking and density formation is visible. The comments accompanying the photo suggest that the grout hole can gradually become blocked because of relatively quiet conditions in it, and that the pockets visible are part of channels which remain active because the concentrated flow in them is forced to be turbulent. The comments then go on to suggest that these channels may become blocked before fissures are properly grouted and therefore premature refusal and poor grouting can result. The remedial action suggested is to drill the hole out and start again. The problem is ascribed to the use of "unstable" settling grouts rather than use of those where bentonite is added with the purpose of reducing settling.

This present paper explains the formation of the channels or pockets differently. As will have been seen from earlier in the paper, their formation can be the result of lack of bleeding off. Bentonite does not eliminate this. The water and thin grout should have been bled out of the top of the hole and replaced with fresh grout. There is no need to add bentonite to achieve quality grouting. "Unstable" grouts can readily produce it if correct technology is used.

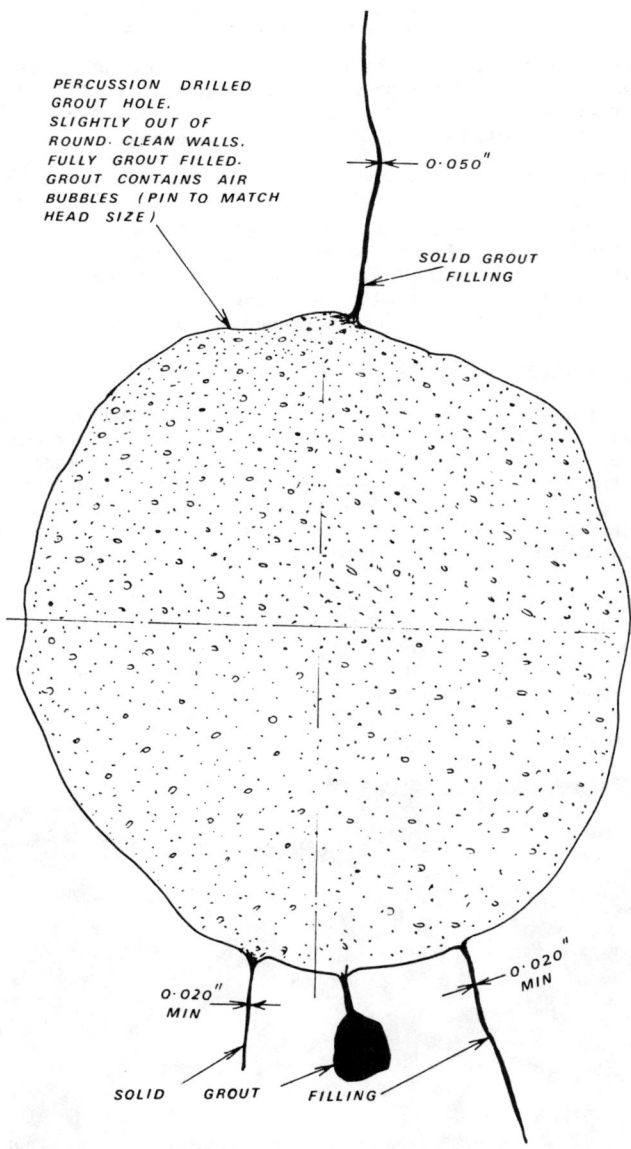

FIG. 10. Observed penetrations and nature of filling in an adequately bled grout hole.

EXAMPLE OF BLEED VOIDS IN POST-TENSIONING GROUTING

The photo referred to in (1) and the bleed voids shown in Figure 9 bear a close resemblance to the large voids and associated chalky grout found in poorly mixed and placed grout surrounding post-tensioning cables in Hume Dam, Australia - described in (4).

EXAMPLE OF PROPERLY BLED ROLE

By contrast, an example of a perfectly grouted hole is sketched at Fig. 10. This sketch is from a borehole periscope observation in a hole which cut a completed grout hole on the skew. The grout hole had been adequately bled and a high speed Colcrete grout mixer had been used. The hole forms part of the grout curtain at Glenlyon Dam, Queensland, Australia.

PENETRATION TESTS BETWEEN SLABS OF CONCRETE

The behaviour of grout during injections has been studied by a number of organisations. Tests have mostly taken the form of penetration in horizontal cracks, some between slabs of concrete, others between transparent sheets. Such tests represent the hardest type of crack to grout because of the inability of bleed water to escape and because of the horizontal orientation. Results should therefore be viewed as somewhat overly pessimistic in relation to most natural cracking systems.

Well known tests of this type are those of the U.S. Corps of Engineers (8) carried out in the early 1950s. Subsequent tests were carried out in 1962 by the writer's organisation, known as the Water Conservation and Irrigation Commission at that time. These tests have been briefly mentioned in (5)(6) but are reported in some detail for the first time in this present paper, by request. Previous publication seemed unwarranted in view of similarity with results from the Corps of Engineers.

FIG. 11   Set-up for tests between slabs of concrete. Grout injected at centre of photo

CEMENT GROUTING

These tests were carried out between pairs of reinforced concrete slabs each 3' x 3' x 2" (0.9 mm x 0.9 mm x 50 mm), securely braced against deflection and bolted and clamped together with shims between the slabs to provide test cracks of various sizes. The cracks were tapered with zero opening one side and either 1/16 inch (1.5 mm) or 1/8 inch (3 mm) the other side. At the centre of the upper slab, grout was applied through standpipe fittings identical with those on regular grout holes. The grout was thus able to travel radially. After it had hardened the slabs were taken apart and the nature of the grout film was observed and photographed. The test slabs are still in existence.

Fig. 11 shows the test setup. Grout was allowed to flow freely out from all four edges of the 0 to 1/16 inch cracks (if it reached them), except in test No. 1. However, a different arrangement was necessary with the 0 to 1/8 inch cracks because of the excessively free flow which would otherwise occur. Therefore foam rubber was used between these slabs near their edges wherever the crack widened out beyond 1/16 inch. This restriction affected radial penetration in these wider cracks and should be kept in mind when reviewing the results. There could, however, occur free bleeding upwards around the bolt holes located near the corner of the slabs.

Features tested were the effects of:

(a) various w:c ratios;

(b) paddle mixing compared with high speed mixers;

(c) pumping with a piston pump compared with a Mono (Moyno) pump;

(d) thickening the mix during grouting;

(e) multiple applications of grout;

(f) roughness of crack walls;

(g) size of crack;

(h) additives.

Prior to grouting, the cracks were saturated with water and this water was drained off except in tests 20 and 21. Grout injection pressure was 30 p.s.i. (2 bars) for the first 15 minutes usually and was then increased to 50 p.s.i. (3.5 bars) and kept at this until 30 minutes after refusal. However, test 12 was kept at 30 p.s.i. throughout because of its very high take.

Most of the tests included attempts to fill bleed voids by giving one or more applications additional to the first at intervals of not less than 24 hours.

All tests were carried out to absolute refusal and pressure was then held on the grout for half an hour.

Table 1 summarises the testing variations in terms of these features. The high speed mixer used was a Colcrete mixer made by Colcrete Ltd., England. The paddle mixer was locally improvised with a vertical spindle, having blades rotating at about 100 revs per minute. The Mono pump was a standard CD 68 unit supplied by Mono Ltd. Australia. The piston pump was a duplex Gardner-Denver, air-operated. In some tests as indicated in the table, the mix was thickened during the application to simulate actual rock grout procedures.

Three tests (Nos. 14, 15 and 16) used additives then marketed in Australia as GF1 and GF2.

Tests 20 and 21 simulated grouting against flowing groundwater pressure, by passing water at 15 ft. head through the crack before, during and after the grouting.

Cement takes are shown in Table 2 in bags or sacks of 94 lbs size, approximately 1 cu. ft. loose volume of cement in each. In some tests much of this cement left the test slabs at the perimeter or through the bolt holes acting as vents. Even though this was happening grouting was continued until refusal was reached in order to simulate as far as possible conditions around a grout hole from which grout is penetrating to a distance beyond the 1.5' (0.5 m) limit represented by the edge of the test slabs. A disadvantage of this method of permitting grout to escape is that cement takes reflect escape conditions rather than being indicative of crack filling considerations. The cement takes in Table 2 and the refusal times shown there should, therefore, not be regarded by themselves, as giving indicative comparisons between the various factors.

Results

The dominant feature of all tests was the degree of unfilled crack where bleed water had been trapped. Photos of examples of this and of other features of the testing are in Figures 12 to 15. In the photos of Figs. 12 and 13, the lighter coloured parts are grout, and the darker parts are empty crack. In Fig. 14, the occasional meanders reaching to roof of the crack are highlighted in white.

Because the prime purpose of grouting is to effectively fill cracks, an important part of the testing was the assessment of the degree to which this was achieved. In some cases grout did not reach to the roof of the crack. Therefore assessment of the degree of filling was made in three ways:

(a) the percentage of the crack area grouted fully up to the top;

(b) the percentage grouted nearly to the top. Some tests showed bleeding above the grout which prevented 100% filling. An arbitrary filling of 95% was used when assessing the percentage area where this happened;

(c) the overall average filling of the crack in terms of the volume of area grouted versus the original open volume of the crack.

TABLE 1 - SUMMARY OF TEST FEATURES

TEST NO.	CRACK SIZE	START-ING W:C	MIXING	PUMPING	THICKENED TO	OTHER FEATURES
1	ZERO TO 1/16"	8:1	PADDLE	PISTON	–	PERIMETER CAULKED
1A	"	"	"	"	–	
1B	"	"	HIGH SPEED	MONO	–	
2	"	5:1	PADDLE	PISTON	–	
3	"	"	HIGH SPEED	MONO	–	ROUGHENED SLAB
4	"	3:1	PADDLE	PISTON	–	
5	"	"	HIGH SPEED	MONO	–	
6	"	1:1	PADDLE	PISTON	–	
7	"	"	HIGH SPEED	MONO	–	
8	ZERO TO 1/8"	8:1	PADDLE	PISTON	–	
9	"	5:1	"	"	3:1	
10	"	3:1	"	"	–	
11	"	"	HIGH SPEED	MONO	–	
12	"	1:1	PADDLE	PISTON	0.8:1	
13	"	"	HIGH SPEED	MONO	–	
14	ZERO TO 1/16"	"	"	"	–	ADDITIVE GF1 USED
15	"	"	"	"	–	ADDITIVE GF2 USED
16	"	"	"	"	–	GF1 AND GF2 USED
17	ZERO TO 1/8"	"	"	"	–	
18	ZERO TO 1/16"	5:1	PADDLE	PISTON	3:1 2:1 1:1	
19	ZERO TO 1/8"		HIGH SPEED	MONO	3:1 2:1 1:1	
20	ZERO TO 1/16"	"	"	"	–	WATER FLOWING IN CRACK AT 15 FT HEAD
21	"	3:1	"	"	–	"
22	"	VARIOUS	"	"	–	

TABLE 2 — CEMENT TAKES AND TIME TO REFUSAL

TEST NO.	CEMENT TAKEN IN EACH APPLICATION - BAGS			TIME TAKEN TO REACH REFUSAL - MINUTES		
	1ST	2ND	3RD	1ST	2ND	3RD
1	0.04	3.47	0.01	15	45	15
1A	2.15	0.01		60	15	
1B	1.32	0.01		15	15	
2	0.31	0.02		30	15	
3	2.46	0.02		30	15	
4	16.43	0.03		45	15	
5	0.40	0.03		15	15	
6	0.20	0.07		15	15	
7	0.20	0.07		15	15	
8	1.42	0.02		30	15	
9	13.05	0.02		90	15	
10	7.34	0.03		45	15	
11	2.88	0.03		30	15	
12	117.88	0.08		210	15	
13	11.46	0.07		30	15	
14	0.47			15		
15	1.07			30		
16	0.40			15		
17	9.73			15		
18	1.45			45		
19	61.13			240		
20	0.40			15		
21	0.46			15		
22	SEE TEXT					

CEMENT GROUTING

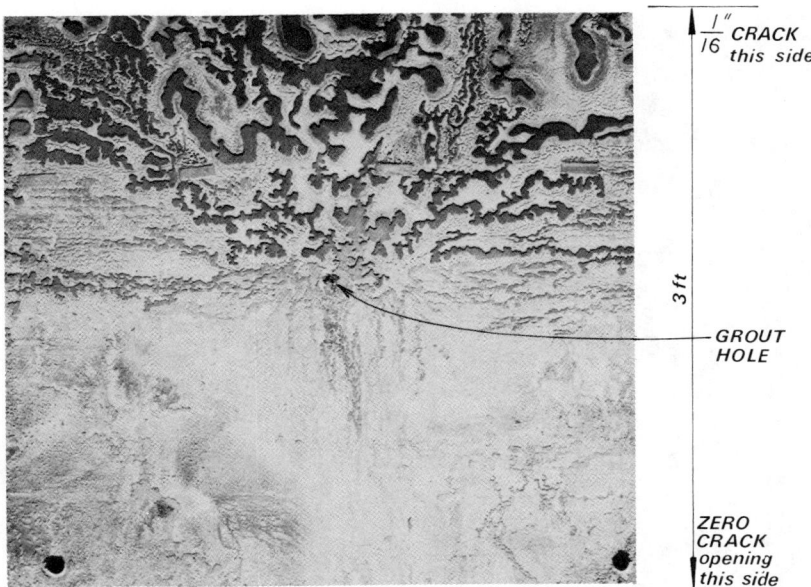

Test No. 4. 3:1 Mix. Paddle mixer, piston pump. Grout able to escape freely from all sides.

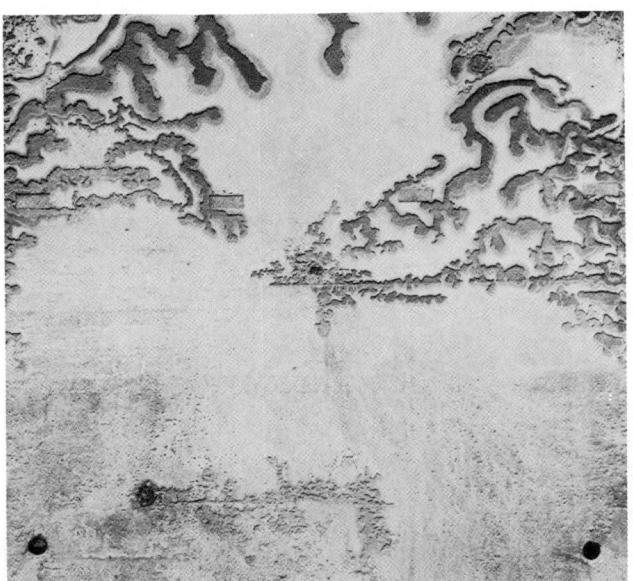

Note better filling than in No. 4, but many unfilled bleed paths.

Test No. 5. Same as for Test No. 4 except that high speed mixer, mono pump used.

Fig. 12.

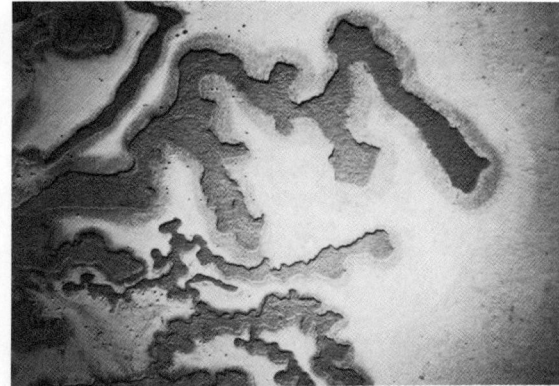

Test No. 5.
Enlargement of top left-hand corner (see previous page).
W:C = 3:1.

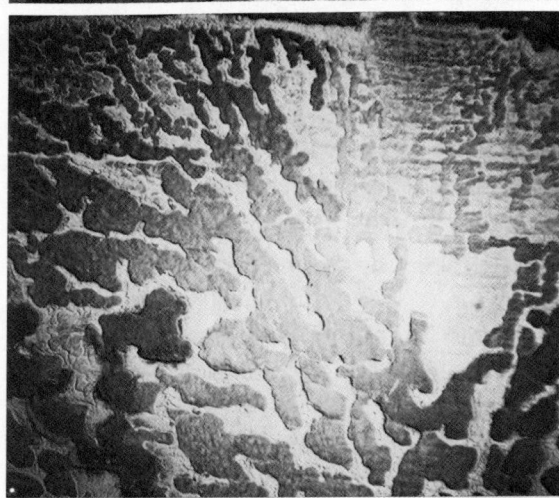

Test No. 1.
Enlargement of top left-hand corner.
W:C = 8:1.

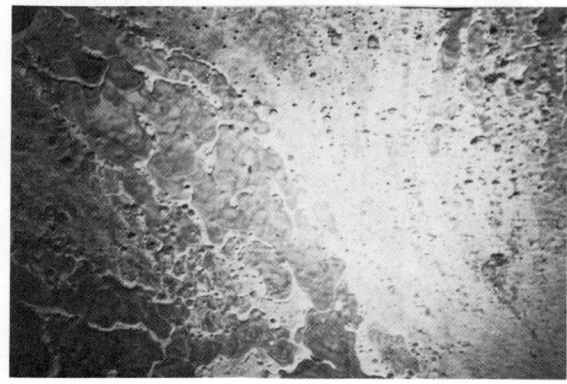

Test No. 2.
Enlargement of top left-hand corner.
W:C = 5:1.

FIG. 13.

CEMENT GROUTING

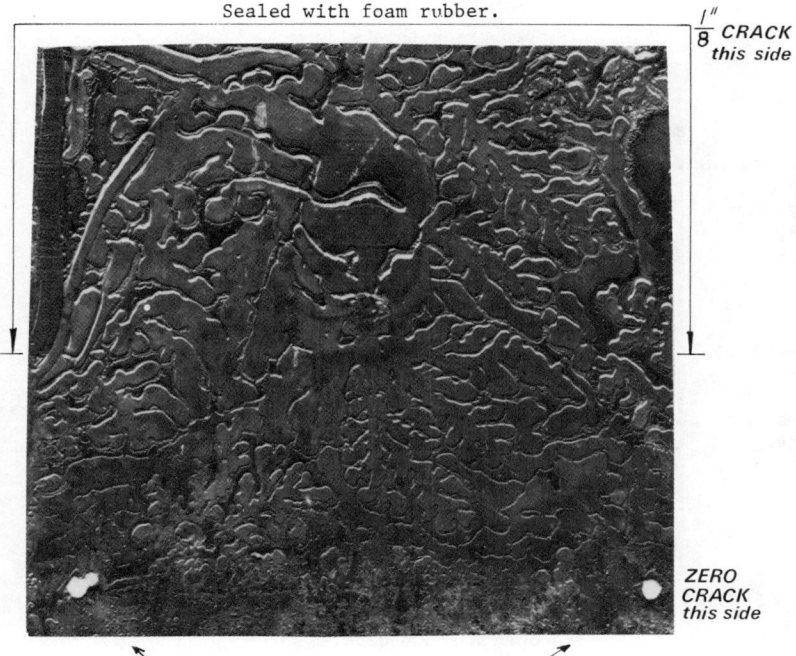

Sealed with foam rubber.

$\frac{1}{8}''$ CRACK this side

ZERO CRACK this side

Venting permitted through bolt holes (at each corner).

Enlargement of top left-hand corner.

Test No. 11. W:C = 3:1
High speed mixer. Mono pump.

FIG. 14.

$\frac{1"}{16}$ open crack this side

Note differing nature of penetration fringe at varying crack width.

Also note ridges produced by rough slab surface.

Radial distribution of grout from grout hole.

Zero opening this side.

Approximate radius of penetration limit is 1 ft.

Test No. 6. W:C = 1:1.
Paddle mixer. Piston pump.

Example of inadequate penetration in fine crack because of use of unsuitably thick mix. Also shows nature of grout at limit of penetration.

FIG. 15.

CEMENT GROUTING

These estimates were made visually by three engineers including the writer. They are subjective and as can be seen from the photos, very difficult to form. Tables 3 and 4 list them together with some comments on visual observations. Because the cracks were tapered the estimates are given separately for the wider and narrower side of each.

Some of the percentage filling results shown on Tables 3 and 4 are plotted on Figures 16 to 19 graphically against the various w:c ratios. Figures 16 to 17 relate to the tests in the 0 - 1/16 inch sized cracks and are for the narrow and wider sides of the crack respectively. Figures 10 and 19 are for the 0 - 1/8 inch crack. Average curves are drawn as a continuous line in the cases of tests using high speed mixing and Mono pumping, and as a dashed line for paddle mixing and piston pumping tests. Tests 1 - 13 and 17 are shown. Omitted are the other tests because of their special variations. Attention is drawn to the fact that only single tests were made for all but one of the factors plotted. This exception was in the case of tests 13 and 17 which were identical in nature. As a measure of the tolerance with which the other results should be viewed it should be noted that tests 13 and 17 gave:

- identical estimates of percent of volume grouted in both narrow and wide sides of the crack

- identical estimates of filling to roof of crack in the narrow side, and 25% and 15% in the wider side

- 25% and 40%, and 60% and 45% respectively in the narrow and wide sides in the assessment of 95% filling.

Trends discernible from the graphs and tables and information from the testing are:

(a) In the narrow side of the 0 - 1/16 inch crack, 3:1 mix gave optimum filling.

(b) In all cracks wider than this, 1:1 gave optimum filling.

(c) In both narrow and wide sides of the 0 - 1/16 inch crack high speed mixing and Mono pumping produced, in general, better filling than paddle mixing and piston pumping for grouts thicker than 5:1. The reverse was partly the case with 8:1 grout.

(d) In the 0 - 1/8 inch cracks, where high speed mixing and mono pumping were only carried out at the thicker mixes, they produced lesser filling possibly than paddle mixing and piston pumping. However the comparison might not be valid in the case of the 1:1 mix where test 12, using paddle mixing, included thickening to 0.8:1 which the others in the comparison did not.

(e) 1:1 mix in the 0 - 1/16 inch crack was obviously too thick for effective penetration and produced a radius of penetration of only about 1 ft. (Figure 15). The nature of the grout filling at the fringe of penetration reflected the crack opening size and the photo shows this.

TABLE 3 - DEGREE OF FILLING ACHIEVED - NARROWER SIDE

TEST NO.	% OF AREA GROUTED UP TO ROOF OF CRACK (a)	% OF AREA GROUTED TO WITHIN 95% OF HEIGHT (b) INCLUDES (a)	% OF VOLUME OF CRACK GROUTED	COMMENTS
1	40	90	90	
1A	50	50	70	SOFT POWDER ON SURFACE
1B	5	5	30	SOFT POWDER ON SURFACE
2	5	40	50	SOFT POWDER ON SURFACE
3	5	5	30	MUCH WEAK, SOFT POWDER
4	15	40	50	
5	10	90	90	
6	25	25	30	DID NOT PENETRATE OVER WHOLE CRACK
7	40	40	60	EXCELLENT BOND TO UPPER SLAB
8	5	5	30	
9	5	40	60	
10	5	20	40	
11	5	5	15	
12	50	95	95	
13	25	25	30	
14	75	75	85	
15	75	75	85	
16	75	75	85	
17	25	40	80	
18	30	40	50	
19	30	70	70	
20	MOSTLY WASHED AWAY			
21	MOSTLY WASHED AWAY			
22	PURPOSELY WASHED AWAY			

TABLE 4 – DEGREE OF FILLING ACHIEVED – WIDER SIDE

TEST NO.	% OF AREA GROUTED UP TO ROOF OF CRACK (a)	% OF AREA GROUTED TO WITHIN 95% OF HEIGHT (b) INCLUDES (a)	% OF VOLUME OF CRACK GROUTED	COMMENTS
1	10	10	40	
1A	15	15	20	SOFT POWDER ON SURFACE
1B	20	20	25	SOFT POWDER ON SURFACE
2	5	10	20	SOFT POWDER ON SURFACE
3	10	15	20	MUCH WEAK, SOFT POWDER
4	30	35	50	
5	15	75	80	
6	40	60	60	DID NOT PENETRATE OVER WHOLE CRACK
7	70	70	85	
8	5	5	40	
9	5	40	60	
10	15	30	60	
11	10	10	20	
12	10	95	95	
13	25	60	75	
14	60	60	75	
15	60	60	75	
16	60	60	75	
17	15	45	75	
18	35	35	40	
19	75	85	85	
20	MOSTLY WASHED AWAY			
21	MOSTLY WASHED AWAY			
22	PURPOSELY WASHED AWAY			

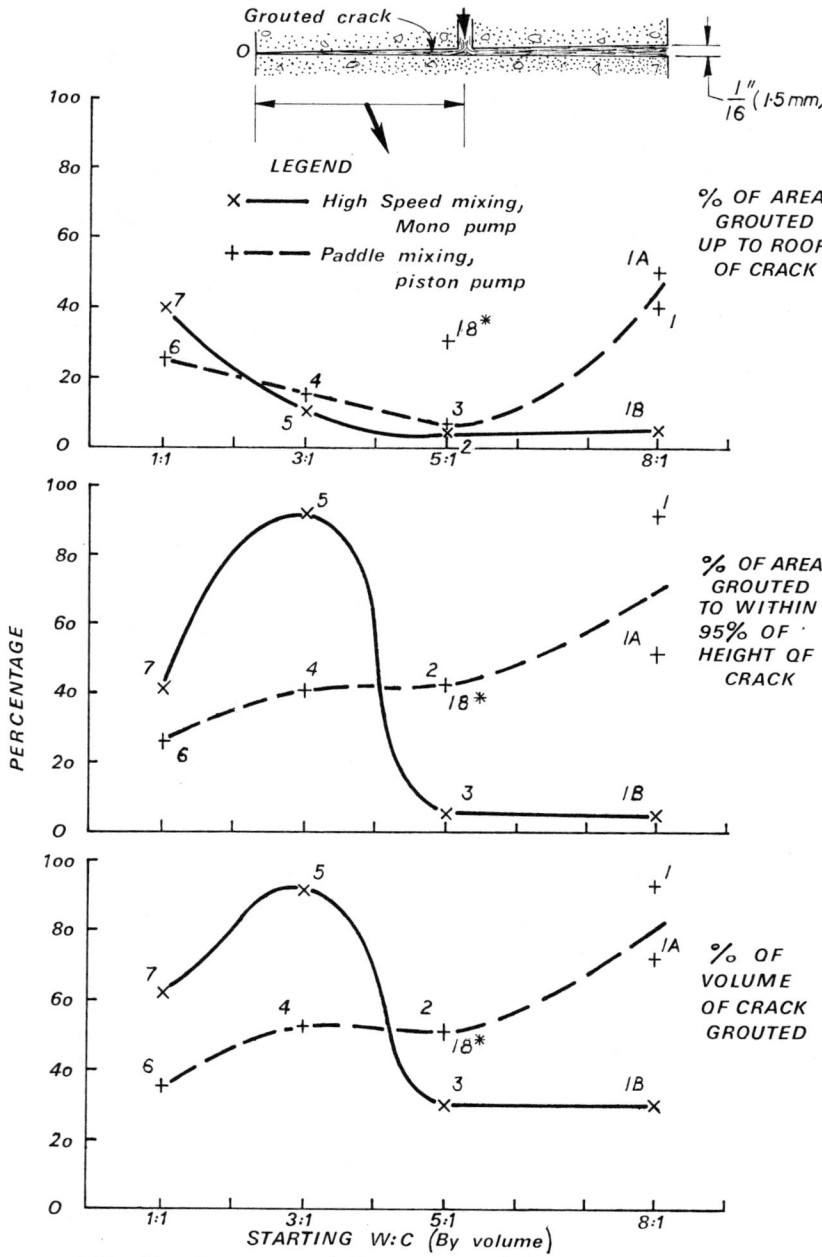

FIG. 16. Percentage filling, 0 to 1/16 inch crack, narrow side.

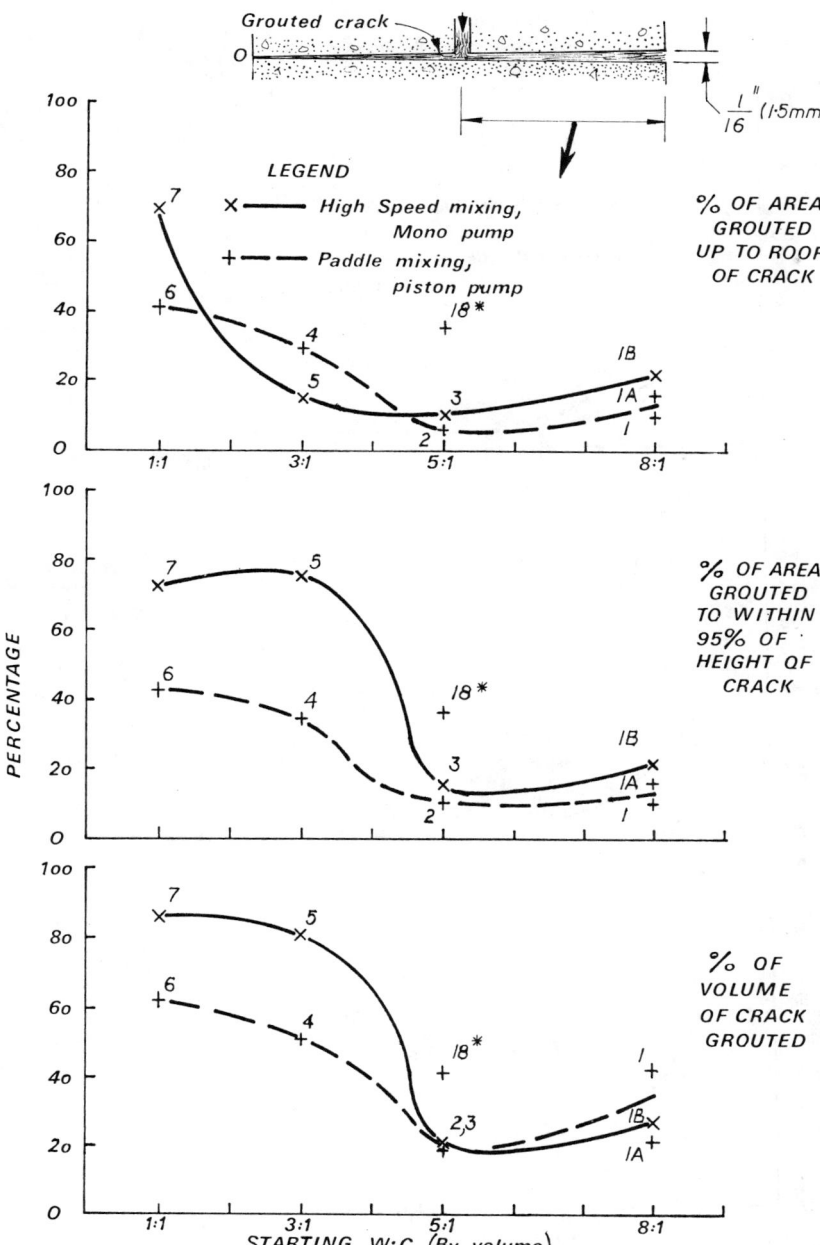

FIG. 17. Percentage filling, 0 to 1/16 inch crack, wider side.

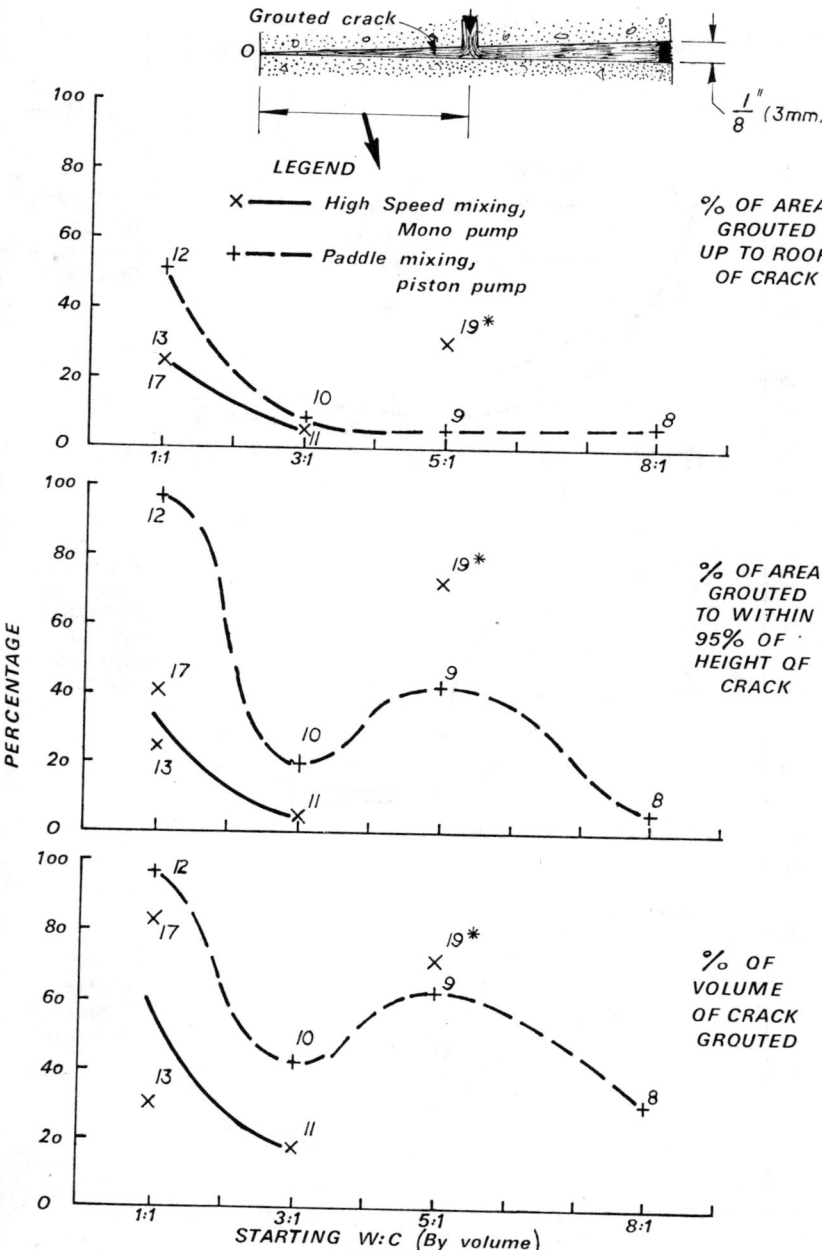

FIG. 18. Percentage filling, 0 to 1/8 inch crack, narrow side.

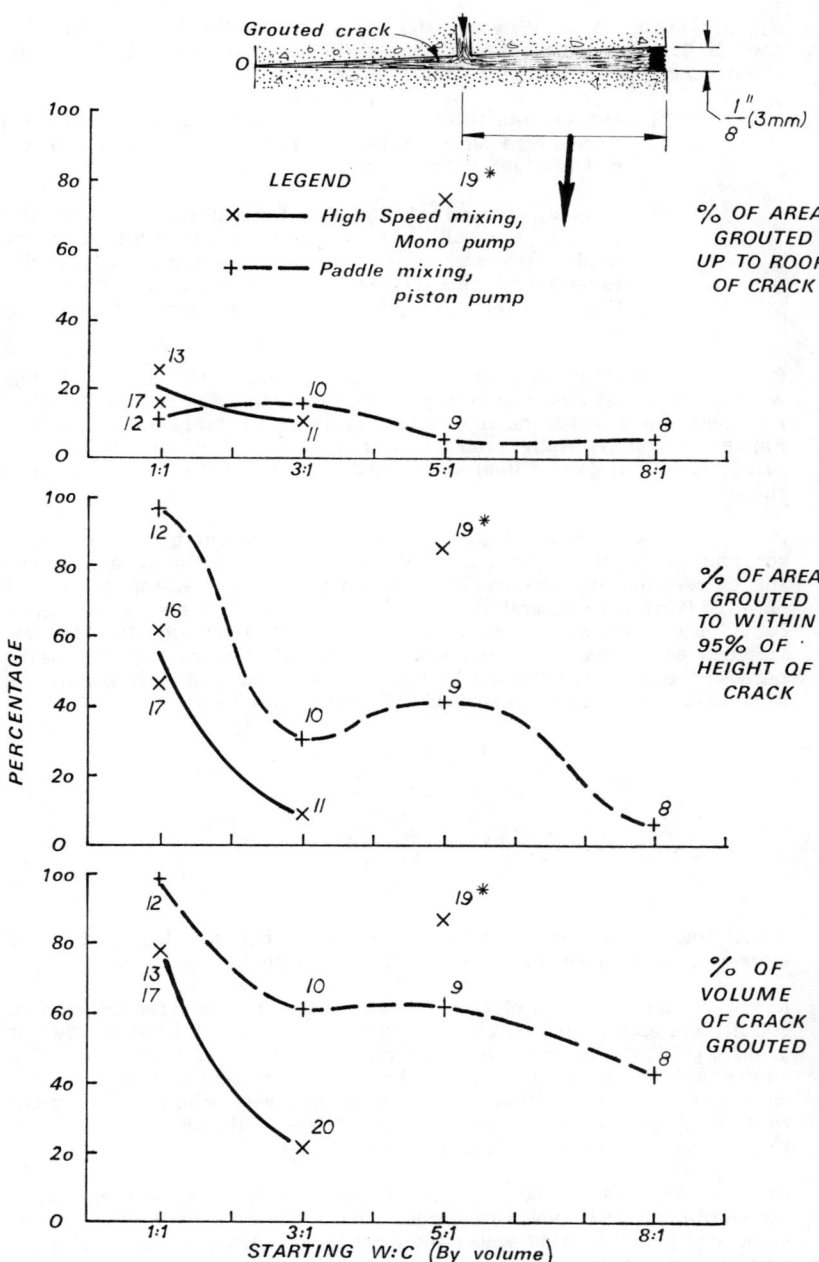

FIG. 19. Percentage filling, 0 to 1/8 inch crack, wider side.

(f) Tests 18 and 19, where the mix was gradually thickened from 5:1 down to 1:1 during the grouting, showed (indicated with an asterisk on the graphs) in comparison with non-thickened tests:

    (i) negligible difference in the 0 - 1/16 crack narrow side and some improvement in the wider side. This test no. 18 was paddle mixed and piston pumped;

    (ii) marked improvement in the 0 - 1/8 inch crack on both sides (Test 19). Results were some of the best obtained in the whole series of tests and were generally only partly exceeded by test 12, which was likewise finished with a thickened mix. Test 19 was high speed mixed and mono pumped.

(g) Artificially rough crack surfaces were made for test 3 and some were unintentionally formed in test 6 and possibly other slabs. The effect of roughness was found to be similar to that shown in the photo of test 6 at Figure 15 and generally produced a ridging effect where the blocked passages and the grout filling were found to be orientated along the rough ridging.

(h) Because some degree of ridging was inevitably produced by screeding inequalities on the slab surfaces, test 22 attempted to overcome this by re-using the same pair of slabs a number of times. Mixes from 8:1 down to 0.8:1 were separately injected between each, the slabs each time being taken apart and the previously injected grout washed off. This was intended as a measure of the cement take but was not very successful because the slabs are thought to have not been provided with exactly the same crack width at each reassembly. Cement takes were:

```
2.99 bags with  8:1 mix
1.20  "     "   5:1  "
2.56  "     "   4:1  "
1.17  "     "   3:1  "
2.20  "     "   2:1  "
2.47  "     "   1:1  "
1.85  "     " 0.8:1  "
```

The thinner mixes tended of course to lose grout out the edges of the slabs more easily and for longer times than the thicker ones.

(i) The simulation of grouting against flowing groundwater, tests 20 and 21, was able to successfully inject grout at 30 p.s.i. (2 bars) and then at 50 p.s.i. (3.5 bars) against a water pressure of 15 ft. head (6.5 p.s.i. approx.) and refusal was obtainable within 15 minutes. Pressure was then held for 30 minutes. However within a few minutes of release of pressure the grout washed out and was almost entirely lost. Mixes used were 5:1 and 3:1. A thicker mix would probably have been more successful.

(j) Tests 14, 15 and 16, with additives, showed improved crack filling but substantial bleed voids remained. The additives were claimed by the supplier to be penetrating agents and were apparently of the lignosulphonate type. However details are not now available.

CEMENT GROUTING

(k) Powdery, weak grout resulted in most of the tests using 8:1 grout. The durability of such grout in flowing groundwater conditions is very suspect. So also is the weak, whitish grout produced along the sides of bleed passages with the thicker mixes. The only grout considered reliably proof against such deterioration is that which filled the crack solidly to the roof, such as in much of tests 5, 12, 19.

EXAMPLES OF APPLICATION OF THE FOREGOING PRINCIPLES

The concept that grout should be as thick as practicable, consistently with reasonable penetration, and that grout holes should be regularly bled, is supported by the following comparisons between grouting projects which have proved to be durable and successful, and some which might not be.

Grouting suffers from the inherent difficulty that the completed work can rarely be actually seen. Evaluation of its success (or otherwise) is therefore dependent on indirect and perhaps inconclusive performance monitoring of various types. With this in mind the following case histories have been grouped as far as possible as "successful" operations or as "not successful as was intended". Note the inability to give a definite "success" or "failure" verdict to grouting, usually.

In general the "successful" jobs were grouted with 3:1 grout or thicker. The others used 5:1 or thinner. This is not to say that all jobs grouted with 5:1 or thinner are, therefore, suspect, but is presented as a discerned distinction between the "successful" and the "unsuccessful" curtains in the limited number of cases known to the writer. Other practitioners may be able to instance further cases to widen the scope of the comparison.

"SUCCESSFUL" CASES

(A) COPETON DAM

This is a central earth core fill dam with a maximum height of 325 ft. (100 m) and a length of 5000 ft. (1520 m).

The foundation is mainly coarse grained massive biotite granite with an inclusion of a small mass of fine grained granite. The rock is in tough, near-horizontal sheets and is tectonically stressed. Main joints are relatively widely spaced and are of considerable extent. Surface joints exhibit effects of stress relief and are frequently wide open. From grouting aspects, the foundation could be regarded as an extreme of which the example shown at (D) is the other extreme. The extreme illustrated by Copeton Dam is that of a foundation where grouting is relatively easy, is obviously needed, and has cracks so wide and extensive that they provide adequate penetration with no difficulty. In fact, penetration had to be curbed in some of the long horizontal cracks which would otherwise have conveyed grout to areas remote from the curtain. However, there were in addition fine, short joints which the grouting also had to cater for.

A single row grout curtain was constructed by stage grouting, using split spacing methods, with widely spaced primary holes then successively halving the distance between holes until eventually the desired standard of permeability was achieved in accordance with the principles of (3).

Figure 20 summarises the grouting. It shows a cross section of the dam and curtain to scale and gives data plotted on ordinates for each sequence of closure. The first ordinate deals with primary holes, the second with secondary holes and so on, and on each is the data relating to starting mixes and then to finishing mixes. This distinction between mixes is necesary because frequently the starting mix is thickened during a grout application and in such cases the starting mix is not really indicative of most of the application.

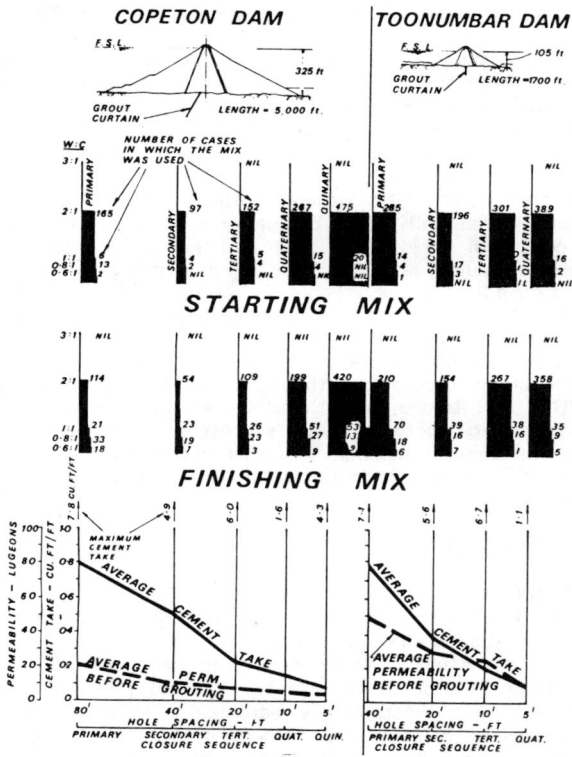

FIG. 20. Summary of water: cement ratios used at Copeton, Toonumbar Dams.

Most applications started with 2:1 grout but 1:1, 0.8:1 and even 0.6:1 were used where water testing immediately prior to grouting indicated the presence of cracks wide enought to accept these mixes readily and the geological investigation indicated absence of fine cracks which would be clogged by such a thick mix. 2:1 was the thinnest grout used.

In primary holes, 2:1 starting mix was used in 165 cases, 1:1 was used six times, 0.8:1 thirteen times and 0.6:1 twice. In secondary holes 97 cases were commenced with 2:1, four with 1:1, and two with 0.8:1. Then as closure progressed, the thicker mixes were needed less, until at the quinary stage (5 ft. or 1.5 m, spacing between holes), 475 of them were started with 2:1, twenty with 1:1 and none with thicker mixes. The total number of cases commenced at the various mixes is listed at Table 5.

TABLE 5

W:C RATIO BY VOLUME	NUMBER OF CASES COMMENCED (FINISHED) WITH THIS W:C RATIO:-			
	DAM			
	COPETON	TOONUMBAR	GLENLYON	CHAFFEY
(1)	(2)	(3)	(4)	(5)
3:1	NIL	NIL	581(376)	371(272)
2:1	1156(896)	1171(989)	16(80)	4(31)
1:1	50(174)	67(182)	117(178)	4(43)
0.8:1	23(115)	10(59)	2(37)	2(35)
0.6:1	2(46)	1(19)	0(45)	NIL

The finishing mix on all these applications was either the same as the starting mix or was thicker. Some applications commenced with 2:1 and were found to require progressive thickening right through to 0.6:1 but the majority of thickening was through less of the range than this. As indicated on Figure 20, 114 cases in primary holes were finished with 2:1 (these had not been thickened during the application), 21 with 1:1, 33 with 0.8:1 and 18 with 0.6:1. Subsequent sequences were in similar proportion. Table 5 also gives in brackets the number of cases finished at the various mixes.

The case numbers quoted are for curtain grouting only, and include holes where surface leaks or connection occurred. They also include any multiple applications given at 24 hour intervals due to the take on the first, or on an earlier application, being sufficiently great to warrant further tightening grouting.

At the foot of Figure 20 is a summary graph showing cement takes at the various closure sequences. The average take on primary holes was 0.79 cu. ft./ft. (= litre/cm.) with a maximum of 7.8 cu. ft./ft. (= litre/cm). This reduced to 0.51 on secondaries with a maximum of 4.9. Subsequent closure showed progressive reductions although the maximum for quinary holes (4.3) showed a higher figure than earlier (1.6) because a number of these holes picked up isolated short joints. This event corroborated the need for closure to 5 ft. in order to produce a thorough curtain to the design standard.

The graph on Figure 20 also shows the average permeability before grouting, at each closure sequence. Before any grouting of primary holes at all, the average permeability was 21 Lugeons. The grouting of primary holes reduced this to 10 Lugeons before secondary grouting. This was reduced to 9 Lugeons by the secondaries, 8 by the tertiaries, 6 by the quaternaries and to an estimated 5 by the quinaries.

The penetration of grout from primary holes at 80 ft. (24 m) spacing exceeded 40 ft. (12 m) in a sufficient number of cases to appreciably reduce the demand in secondary holes.

The grout was mixed at high speed (greater than 1,500 r.p.m.) and was pumped with Mono pumps. Grouting techniques were generally in accordance with (9).

The dam has been storing water since 1974. It shows negligible leakage, typically 0.07 c.f.s. (2.2 l/sec) for full storage conditions, and pore pressures are satisfactory.

(B) TOONUMBAR DAM

This resembles Copeton Dam in cross section approximately but is only 105 ft. (32 m) high and 1700 ft. (518 m) long. On Figure 20 it is drawn to the same scale as Copeton Dam.

The foundation comprises various beds of fine, medium or coarse grained sandstone separated by thin lenses of fine conglomerate shale breccia or shale and laminite. The general strike is normal to the axis of the dam: dip is about 10°.

The rock is relatively weak, particularly in the laminated horizons. Cross bedding dips in variable directions at up to 15° from the true bedding. The beds tend to fracture along cross bedding causing local high angle discontinuity planes. The beds have been fractured by many near-vertical joints and piping of weak material from seams had to be considered when designing the grouting.

From considerations of ease of grouting, the foundation could be regarded as intermediate between examples (A) and (D).

The grouting data is presented here on Figure 20 in the same way as for Copeton Dam. The grouting was all at 2:1 or thicker.

285 of the primary holes were grouted with 2:1 as the starting mix and 14 with 1:1. The joints were too fine to require thicker starting mixes than this apart from 4 cases of 0.8:1 and one case of 0.6:1. This pattern was similar in subsequent sequences. Thickening during grouting resulted in finishing mixes in primary holes of 1:1 in 70 cases, 0.8:1 in 18 and 0.6:1 in six. Table 5 gives total number of cases for each of the starting and finishing mixes.

The graphs at the foot of Figure 20 show the reductions in cement takes and permeability which occurred as a result of closure. The average cement take in primaries of 0.79 cu. ft./ft. (= litres/cm) indicated a ready degree of acceptance of the 2:1 grout in the relatively fine jointing.

Grouting equipment and methods were the same as for Copeton Dam. The dam has been storing water since 1971 and is behaving satisfactorily.

(C) GLENLYON DAM

This is another central core, earth fill dam. Dimensions are shown on Figure 21, where its cross section is drawn to the same scales as those of Figure 20.

Indurated fine grained claystones form the foundation, with thin quartzite bands and frequent calcite and quartz veins. Secondary structural features cause discontinuities such as Breccia zones, shearing and multi-directional jointing.

Overall, the foundation could be regarded as fairly weak structurally. It is intensely jointed in near-vertical and near horizontal patterns, some of which are associated with narrow shear zones. Preliminary grouting was carried out to tighten some exceptionally wide cracks. After that its groutability was about the same as example (B).

3:1 was the thinnest mix used here. The details are on Figure 21 and Table 5. These include the Preliminary grouting. Closure produced the unusual situation where the plot of take-reduction is almost a straight line. Maximum cement takes on primary and secondary holes were unusually high at 37.4 and 19.8 cu. ft./ft. (= litre/cm).

CEMENT GROUTING 71

The pattern of permeability reductions during the closure showed that, on average, tertiary sequence was the first which produced substantial change. Tertiary holes brought the spacing down to 10 ft.; penetration was of the order of 5 ft. therefore the subsequent (quaternary) holes encountered lower permeabilities than earlier holes had. Even so, the permeability prior to quaternary grouting was 15 Lugeons average and therefore the quaternary grouting was essential to bring the permeability down to the desired standard.

Equipment and methods were the same as for Copeton and Toonumbar Dams. The dam is behaving satisfactorily. It commenced storing water in 1977, and filled in 1984.

(D) CHAFFEY DAM

This dam is included because it illustrates a somewhat extreme situation where penetrations were untypically short because of fine cracking.

As Figure 21 indicates, closure produced little effect until after tertiary sequence was reached, but because the aim of this particular grouting was somewhat unusual in that it did not have the purpose of producing a low permeability curtain but had the purpose of dealing with significant cracks only, very little grouting was carried on beyond the tertiary sequence, which was at 10 ft. (3 m) spacing. The foundation had an average permeability of 19 lugeons before grouting of primary holes. The increase to 24 lugeons measured in quaternary holes before grouting, as shown on Figure 21, does not result from hydraulic fracturing during previous sequences but is merely an example of the vicissitudes produced by the statistical averaging of grout takes. Averaging is never a wise approach considering the wide ranges experienced in grout takes, but is used here of necessity for illustrative purposes. Quaternary and quinary holes were used only where takes in tertiaries warranted them. The overall permeability on completion of the grouting is, of course, rather variable in such a situation. It appears to be of the order of 15 lugeons which is much the same as before grouting and raises the issues as to why do the grouting at all! The purpose was, as indicated above, to deal with discrete significant open joints. The criterion for continuation of closure sequences was based on cement takes rather than on permeability and was set at 0.5 cubic feet of cement per foot.

The geology of the foundation can be simplified into bars of tough, almost tight, vertically dipping massive jasper, interspersed with areas of weak siltstone and chert. The strike of beds approximates the axis direction. Seepage emerges through deep alluvial beds without known hazards to the stability of the dam and seepage losses of several c.f.s. are tolerable.

The range of penetration from many of the grout holes possibly averaged only a foot or two, hence in the sections where closure ceased at 10 ft. spacing there conceivably remain ungrouted cracks, possibly wide enough to warrant grouting but not accessed from the grout holes. A closer spacing of final holes would be necessary to thoroughly grout all groutable cracks.

The dam has filled and is behaving satisfactorily. Seepage is unable to be measured, but is slight.

From the grout technology point of view this foundation is of the type which might suggest that use of very thin mixes would have been better than those actually used, or to some European practitioners it might suggest the use of displacement high pressure techniques to burst open long lengths of cracks and then gain wide access from few holes. When deciding the w:c.

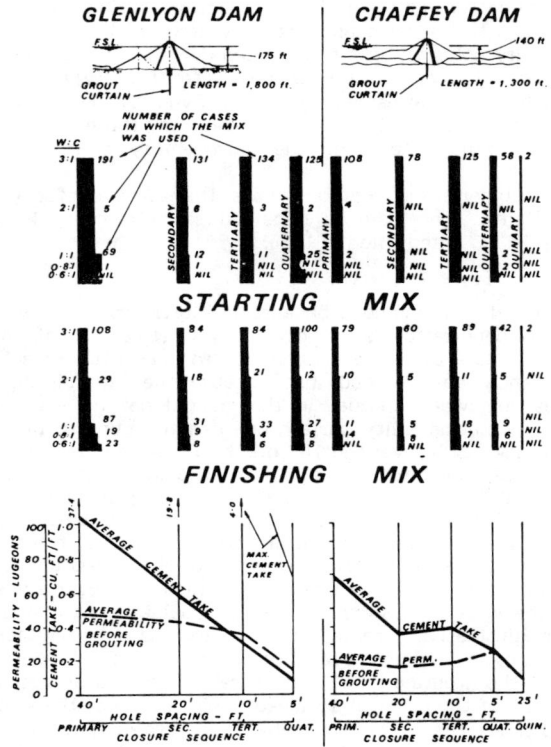

FIG. 21. Summary of water:cement ratios used at Glenlyon, Chaffey Dams.

ratio to be used, the durability of the completed work was regarded as paramount, and hence 3:1 and thicker mixes were used for reasons described. Realistic improvements in penetration would have required the use of mixes of about 10:1 or thinner; the use of these is regarded as a wasted effort in terms of long term efficacy. The displacement method is regarded as somewhat hazardous, disrupting as it does the natural bonding in near tight rock (which would not have warranted grouting if undisturbed) and opening up, in this site, numerous near vertical cracks along the axis, most of which would leak grout embarrassingly to the surface. Hence the grouting was carried out by normal methods, using 3:1 or thicker.

(E) WARRAGAMBA DAM

This is a concrete gravity dam, the highest in the southern hemisphere with a height of 450 ft (137 m) and a crest length of 1152 ft. (351 m). The foundation is horizontally bedded sandstone mainly, exhibiting near-vertical and horizontal cracking with openings varying widely from almost tight to several inches. The larger openings were due to tectonic stress relief manifesting itself in rebound of the valley floor and block separation of the canyon walls before and during grouting.

3:1 or thicker was used for all grouting. This was a most extensive grouting operation comprising several curtains and including shallow blanket grouting of the whole foundation. The total length of holes was 1,117,000 ft. (340 000 m) and the total cement injected was 293,850 cu. ft. (8 320 000 litres). The average take was 0.26 cu. ft. of cement/ft (= litre/cm).

The grouting was carried out between 1953 and 1960 using paddle mixers and reciprocating pumps of various types, although towards the end of the job, the deficiencies of this equipment led to experiments with better equipment such as subsequently has been generally adopted elsewhere. An unusually great number of grout holes was necessary in a number of locations in order to compensate for these equipment deficiencies.

The dam has been storing water for 20 years approximately and is performing satisfactorily.

CASES WHERE GROUTING WAS NOT AS SUCCESSFUL AS WAS INTENDED

(A) BURRINJUCK DAM

This is an old concrete gravity dam which leaked so much through the concrete that it was grouted over the period 1936 to 1948. The grouting comprised a 3 row curtain drilled from the crest through the dam and into the foundation. The old concrete was of poor quality and cracks in it took most of the grout. In the initial curtain row, the average cement take was 0.18 cu. ft./ft. (= litres/cm) with a maximum of 8.1 (in concrete). The later two rows took less.

Water:cement ratios were 5.5:1 to 30:1 with an average of 12:1.

The grout curtain in the rock is being gradually leached away by seepage which then emerges in tunnels and deposits the leached calcium in the form of stalagtites and shawls. This grouting appears to have been non-permanent because of the use of excessively thin grout. However, remedial work is not necessary in this case because the foundation was really too tight to need grouting in the first place, and because the tunnels provide adequate drainage without assistance from grouting.

(B) KEEPIT DAM

This is a modern concrete gravity dam with conventional single row grout curtain and drainage hole arrangements. The foundation is finely jointed conglomerate and tuff and the curtain grouting was commenced with thin grouts, such as 8:1. However, it was found that this grouting was ineffective in reducing uplift pressures, probably partly because seepage passing through the foundation at the time of grouting washed much of the grout away before initial set occurred. Tests indicated that initial set for the 8:1 grout could take several days, and even then the resulting product was very weak.

A limited test in situ using 0.8:1 in new holes alongside those using 8:1 average, was able to inject the same small quantities of cement in the cracks as had been injected with 8:1. The initial setting time of 0.8:1 is only a matter of hours and assisted as it is by thixotropy, it resisted removal by seepage.

Recently, regrouting has been found necessary in a vital part of the curtain which has been progressively deteriorating and passing undesirably large quantities of seepage. The original grouting was done in 1961 with 5:1 mix. The regrouting used 3:1 and thicker, and injected considerable quanitities of cement into the deteriorated curtain.

(C) HUME DAM

This is another concrete gravity dam and was constructed initially between 1919 and 1936. By 1954, some concern for uplift pressures then existing, together with modifications for increasing the storage, prompted attempts to construct a grout curtain. The foundation is granite with occasional shears and jointed areas.

A conventional curtain was attempted at various locations along the dam using 20:1 grout. This mix was adopted because, at the time, it was believed by those concerned that grout thicker than this would not penetrate in the very fine joints. A small amount of bentonite was used in the grout for lubrication purposes.

However, after attempts in several short areas it soon became apparent that the grouting was producing no effect on uplift pressures. Some foundation drainage holes were then drilled as an alternative and were so successful in reducing pressures that further grouting was discontinued. Instead, curtains of drainage holes were drilled.

In retrospect, it seems probable that a mix of about 3:1 would have been successful. However, curtain grouting is obviously superfluous at this site, in general, in view of the efficacy of drainage.

(D) BLOWERING DAM

Some parts of the curtain at this dam have disintegrated. The grouting used 5:1 mix mostly and techniques were not the best. It should be noted, however, that some parts remain effective as illustrated at Fig. 4 of (3).

Seepage through the disintegrated grouting and elsewhere is estimated at 11 c.f.s. (300 litres/sec).

SUMMARY

This review of water minimising practices in cement grouting operations:-

(a) Emphasises the desirability of using grouts of w:c=3:1 by volume, or thicker.
(b) Stresses the need to regularly release bleed water and thin grout from active grout holes.
(c) Shows the proneness of grouts to develop bleed passages and voids in conditions where bleed water cannot escape.

REFERENCES

1. CAMBEFORT, H., "The Principles and Applications of Grouting", Quarterly Journal of Engineering Geology, The Geological Society, London, England, Vol. 10, 1977, pp 57-95.

2. HOULSBY, A.C., "Rock Movements During Grouting" Rock Mechanics Symposium, University of Sydney, Australia, Feb 1969, p 116 to 120.

3. HOULSBY, A.C. "Engineering of Grout Curtains to Standards", Journal of the Geotechnical Engineering Division, ASCE, GT9, Sept. 1977, p 953-970.

4. HOULSBY, A.C. "Grouting and Sheathing of Post Tensioning Tendons", Bulletin of Australian National Committee on Large Dams, No. 58, Feb 1981, p 33-35.

5. HOULSBY, A.C. "Cement Grouting for Dams", Proceedings of Conference on Grouting in Geotechnical Engineering, ASCE, New Orleans, 1982, p1-34.

6. HOULSBY, A.C. "Optimum Water:Cement Ratios for Rock Grouting", Proceedings of Conference on Grouting in Geotechnical Engineering, ASCE, New Orleans, 1982, p 317 to 331.

7. HOULSBY, A.C. "A Digest of Typical Cement Grouting Takes" Proceedings of Conference on Grouting in Geotechnical Engineering, ASCE, New Orleans, 1982 p 1000 to 1014.

8. U.S. ARMY CORPS OF ENGINEERS, "Pressure Grouting Fine Fissures", Technical Report No. 6-437, Waterways Experiment Station, Vicksburg, Miss., October, 1956.

9. WATER RESOURCES COMMISSION, N.S.W. "Grouting Manual" Fourth Edition, 1981, W.R.C., North Sydney, N.S.W. Australia, ISBN 0 7240 3874 4 (also available from University of Missouri-Rolla, Civil Engineering Dept.).

USE OF ASPHALT IN TREATMENT OF DAM FOUNDATION LEAKAGE, STEWARTVILLE DAM

Boro Lukajic[1], Grant Smith[2], John Deans[3]

ABSTRACT

This paper presents the recent development and application of a hot asphalt grouting technique used to stop a 5,000 Imperial Gallons Per Minute (IGPM) (22,000 lpm) dam leakage under full reservoir operating head.

Two zones of major leakage were successfully sealed in 1983 and 1984 by simultaneous injection of hot asphalt and conventional cement grouts. This method was necessary in order to deal with the high velocity, turbulent water flow without plugging the adjacent foundation drain. In addition to improving the integrity of the dam foundation, the grouting eliminated water flow within the dam's inspection tunnel which was a safety hazard to personnel and the plant.

While asphalt grouting may have been performed elsewhere under different field conditions, to our knowledge, this is the first instance of its use in Canada.

INTRODUCTION

Stewartville Generation Station, owned and operated by Ontario Hydro, is located on the Madawaska River in Eastern Ontario. The main dam is a concrete gravity structure, 206 feet (63 m) high and 815 feet (248 m) long, Figure 1. An adjoining concrete wingwall on the south abutment is 480 feet (145 m) long. The station was built in 1948 (3 units) and two additional units were added in 1969. Generating capacity is 153,000 kW, with a net head of 150 feet (46 m).

DRAINAGE SYSTEM

The foundation drainage system for the dam is shown in Figure 2. The principal pressure relief system is a foundation drain located 10 feet (3 m) downstream from the upstream face. An inspection

---

1,2   Supervising Geotechnical Engineers, Ontario Hydro, 700 University Avenue, Toronto, Ontario, M5G 1X6

3   Concrete Technologist, Ontario Hydro, 800 Kipling Avenue, Toronto, Ontario, M8Z 5S4

Figure 1. View of the Main Dam

tunnel, 645 feet (197 m) in length, is located above the foundation drain. A series of 6-inch (152 mm) diameter vertical risers connect the tunnel with the drain. The risers serve as a secondary pressure relief system in the event of blockage within the main drain. Water collected in the foundation drain empties into two 12-inch (305 mm) diameter pipes, one from the north end and the second from the south end of the dam. These connect to a single 24-inch (610 mm) diameter pipe which discharges to the tailrace.

FOUNDATION CONDITIONS

Foundation bedrock is predominantly crystalline limestone, which is massive and competent for the most part. Zones of weathered, decomposed micaceous limestone occur along some of the bedding planes and joint systems. Initial foundation preparation was insufficient to treat all of these zones, which are susceptible to erosion by moving water. Some of the geological features in the exposed bedrock foundation are shown in Figure 3. A fault zone which required extensive dental treatment during construction can also be seen in the photograph.

Only limited consolidation and curtain grouting was carried out during construction as shown in Figure 2. It became apparent that some of the zones containing weathered material, particularly near the dam/rock contact, were not adequately treated during the early grouting operations. Leakage from the headpond through the foundation over a period of years eventually washed out and enlarged some of these zones, allowing water to enter the foundation drainage system.

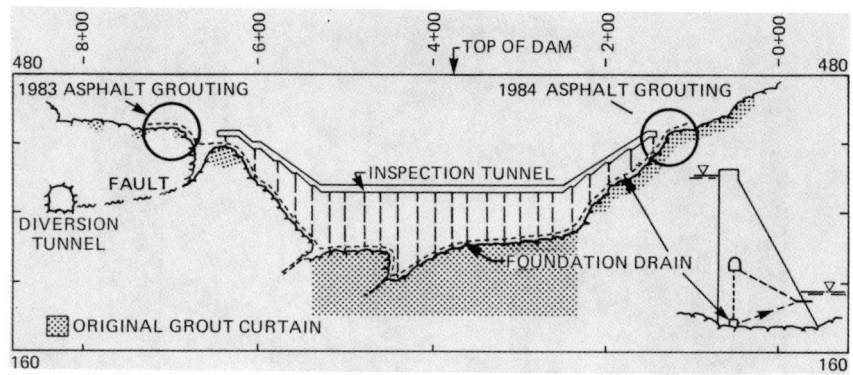

Figure 2. Drainage System

Figure 3. Excavation During Construction - South End

BACKGROUND OF LEAKAGE

Early Leakage Problems

Locations of leakage zones which have required remedial treatment since construction are shown in Figure 4.

Shortly after the reservoir was filled in 1948, leakage occurred along a flat-lying seam beneath the south wingwall. Cement grouting with coarse aggregate filler was performed to stop the leakage.

Leakage into the foundation drain was first observed in 1952. It increased in volume until, in 1959, cinders were dumped in the reservoir close to the upstream face of the dam. This operation temporarily reduced the leakage from an estimated 3000 Imperial Gallons Per Minute (IGPM) (13,638 lpm) to 300 IGPM (1364 lpm).

In the mid 1960's, two units were added to the generating station. During the bedrock excavation for extension of the powerhouse, leakage exceeding 10,000 IGPM (45459 lpm) was encountered beneath the deepest section of the dam, just outside of the zone of the original grout curtain. An extensive cement grouting program was required to stop the leakage, which resulted in costly construction delays. A total of 63,000 bags of cement, 44,000 cubic feet (1246 m^3) of sand and large quantities of fillers were injected into the bedrock. Grouting was difficult as a result of the extremely high velocity water flow, however, the leakage was effectively controlled. Remedial treatment of this problem is described in detail in Reference 3.

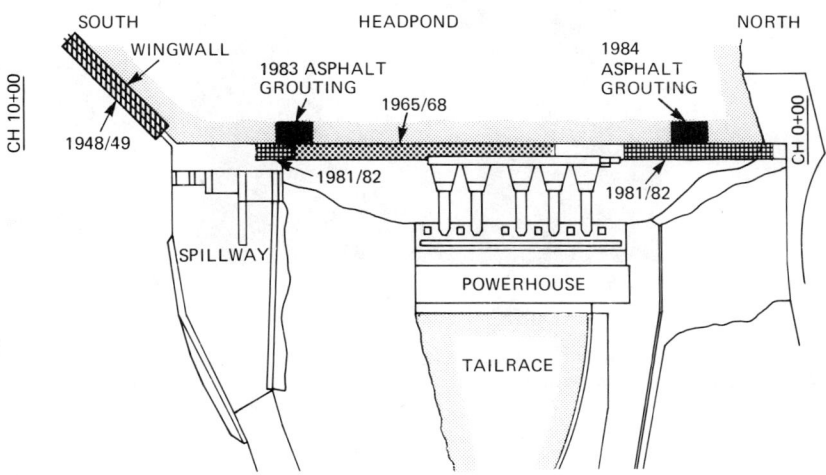

Figure 4. Location of Leakage and Remedial Treatment

Recent Leakage Problem

During the above remedial work, some leaks were not treated with grout. These gradually increased throughout the 1970's. Eventually, the flow exceeded the capacity of the foundation drain resulting in significant discharge into the inspection tunnel mainly along the south stairwell, as seen in Figure 5. Discharge into the inspection tunnel, measured at more than 1000 IGPM (4546 lpm), at times created a safety hazard for maintenance crews. Upstream blanketing was carried out periodically, but reductions in flow that were achieved were only temporary.

Figure 5. Leakage in Inspection Tunnel

Two leakage areas were identified in 1979 by dye testing and diving inspection. One zone was at the south end of the dam next to the spillway. The other zone was near the north abutment, as shown in Figure 4.

Cement grouting was carried out in these areas in 1981/82 to eliminate leakage in the tunnel and to improve the integrity of the foundations. Grouting proved to be no more effective than upstream blanketing in reducing the leakage, as shown in Figure 6. Within a few months of these treatments, leakage noticeably increased.

ASSESSMENT OF REPAIR TECHNIQUES

It was apparent that conventional cement grouting and blanketting techniques were not effective in permanently stopping the leakage.

This prompted a review in 1983 of a wide range of potential remedial treatments. It was concluded that grouting still had the best potential for stopping the flow.

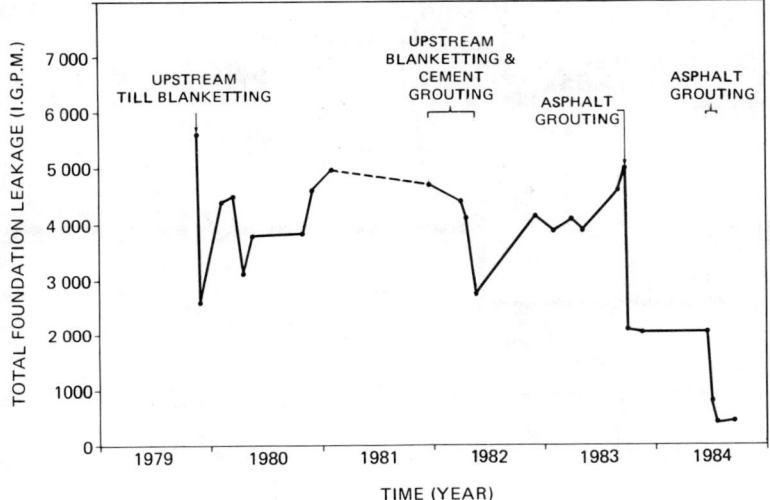

Figure 6. Effect of Remedial Treatment on Leakage

In view of this decision additional field investigations were initiated to define more closely the location and size of the open seams and the velocity of the leakage flow. At the same time, a review was instituted of alternative grout materials.

FIELD INVESTIGATIONS

Grouting operations were preceded by a program of diamond drilling and in situ testing performed in conjunction with the laboratory testing.

As a first step a concrete pad was placed adjacent to the upstream face in the leakage area at the south end of the dam in which to collar the exploratory holes and to confine the grout, Figure 7(a). Several exploratory boreholes were drilled through this pad at the locations shown on the same figure. The holes were located upstream of the dam to permit the injection of grout directly into the seam at some distance from the foundation drain. Several observation holes were also drilled from the top of the dam and from the inspection tunnel.

The north abutment, next to the upstream face, was too steep and irregular for placement of a similar concrete pad. As a result all grout holes were drilled from the top of the dam (close to the upsteam face). Several holes were also drilled within the inspection tunnel to observe the water conditions and monitor grout movement in relation to the drain.

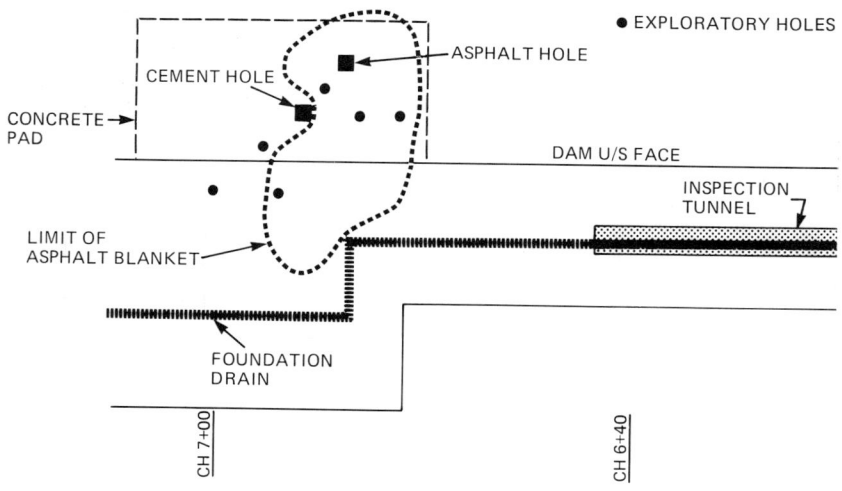

(a) Grout Hole Arrangement

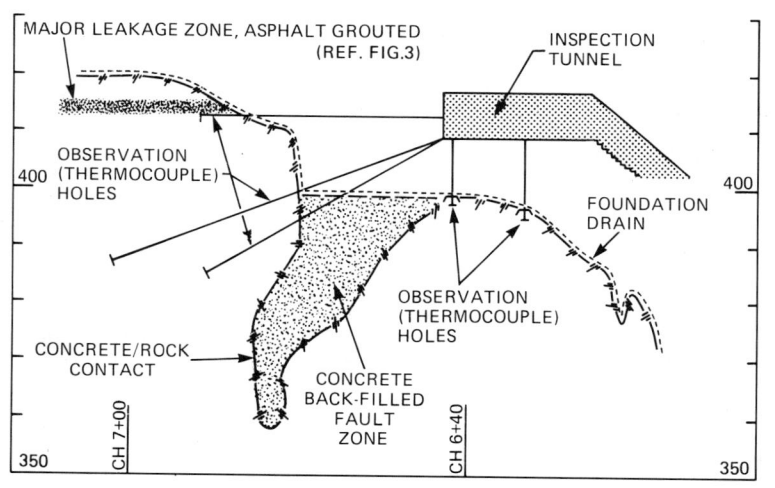

(b) Observation (Thermocouple) Holes

Figure 7. 1983 Asphalt/Cement Grouting - South End

As part of the exploratory program, extensive in situ testing was performed to determine the nature of flow and general foundation conditions. In brief, the tests consisted of:

(a) Downhole television camera viewing to observe the in situ joint conditions as well as water movement within the rock.

(b) Saline tracer tests to establish the rate of flow and the hydraulic communication between the boreholes and the foundation drain.

(c) Dye tracer tests to determine the sources of leakage under the dam and the corresponding points of water inflow into the inspection tunnel.

The test results delineated the leakage area at the south end of the dam as a narrow open bedding plane in the upper part of the bedrock. It was approximately 10 feet (3 m) wide, with up to 8-inch (203 mm) vertical opening. This zone was adjacent to the south side of a concrete backfilled cavity shown on Figures 3 and 7(b).

Leakage beneath the northern portion of the dam (Figures 1 and 2) occurred along the concrete-bedrock contact and through the fractured rock surface. This zone was approximately 15 feet (5 m) wide, with fracture openings in the range of several inches.

LABORATORY EVALUATION OF GROUT MATERIALS

The field investigations revealed the grout would have to perform under the following site conditions:

(a) High rates of water flow in rock

(b) Water/rock temperature approximately $50\pm^\circ F$, ($10\pm^\circ C$)

(c) Fracture sizes in rock in the order of 8 inches (203 mm).

(d) Relatively short distance (up to 25 feet, 8 m) between injection point and the foundation drain, which had to be kept open.

Some of the desirable grout properties were fast set, high viscosity and resistance to erosion. A review of the literature indicated that numerous materials were in use in engineering practice which might meet the above criteria. Asphalt, urethane (water reactive grouts) and cement grouts with accelerating admixtures were selected for laboratory tests to assess their effectiveness in sealing the leakage. A brief evaluation of each grout material is presented as follows:

Cement Grout

From past experience, it was doubtful that cement grouts alone would produce an adequate seal if the grout pumping rate was less than the actual rate of water flow. There was also the potential for large volumes of grout being washed into the drain and possibly plugging it.

It was determined by the laboratory tests that the use of accelerating admixtures and cement grout might be effective, however, because of rapid stiffening this could cause plugging of the grout lines. Since special field handling procedures and set up would be required to prevent premature plugging of the lines this method was abandoned.

Urethane Grouts

It was determined that urethane grouts would be effective in sealing small fractures; however, the combination of large fractures, high gradients and proximity to the box drain would result in little reduction in flow and possible plugging of the drain.

Asphalt Grout Concept

Of the three materials tested it was determined that hot asphalt, simultaneously injected with cement grout had the best potential for stopping the water flow with the least danger of plugging the foundation drain. The concept of simultaneous injection of the two materials is shown schematically in Figure 8.

Hot asphalt was to be injected in the open seams first in order to provide an initial plug to reduce the water flow. This reduction would be achieved by rapid cooling of asphalt as it assumed a globular form when making contact with cold water and subsequently spreading and conforming with the cavity shape. Further pumping of asphalt material into the seam would force the asphalt to mould tightly against the rock surfaces so as to reduce the passage of water.

The injection of cement mix was to be commenced at this point (several minutes later) but through a separate borehole, also intersecting the main seam system. This concurrent injection was to continue as long as the passages were accessible for acceptance of cement grout - to

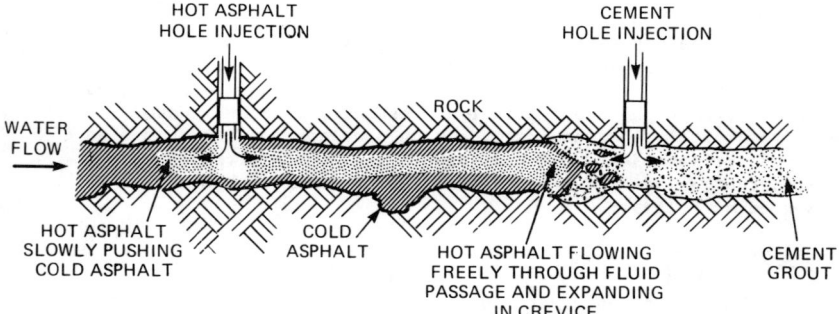

Figure 8. Schematic View of Asphalt/Cement Grouting

ensure that a permanent seal was achieved. It was concluded that asphalt alone could not perform the task as a result of its inability to penetrate small cracks and potential of creeping under hydrostatic head in the larger fractures.

Asphalt Grout Characteristics

The asphalt grout tested in the laboratory and used in the field had the following characteristics:

Type:	CSA Type 3 Roofing Grade - Hard Grade/1/
Softening Point:	205°F (96.5°C)
Flash Point:	527°F (275°C)
Density:	1.04 at 73°F (23°C)
	1.0 at higher temperatures
Minimum Pumping Temperature:	347°F (175°C) (depending on pump capacity)

GROUTING PROGRAM

Equipment Set Up and Testing

The equipment arrangement for asphalt grouting is shown in Figure 9. The major components were standard, readily available, equipment items. The Moyno 3L8 pump was modified by replacing the rotor and stator with stainless steel parts in order to resist the high temperature of the asphalt.

To determine the performance of the heating system as well as the relationship between pumping rate, temperature and pressure, a field trial of the system was conducted prior to grouting of the seam, Figures 10 & 11.

The main conclusion reached from these tests was that insulating both the inner and outer pipes was necessary in order to maintain the desired temperature of 230°F (110°C) within the grout hole. The fibreglass was found to be an adequate insulating material, however, it would have to be protected with a waterproofing material when submerged in water.

The pumping rates of asphalt would likely fluctuate, but would not need to exceed 10 gallons per minute (45 lpm) flow. As it turned out a 4 gallons per minute (18 lpm) flow rate was maintained for about 80 percent of total pumping time, Figure 11.

Asphalt/Cement Grouting Operation

On September 23, 1983 grouting of the leakage zone at the south abutment commenced with injection of hot asphalt through the borehole exhibiting the most open conditions, Figure 7(a). The injection of a sanded cement mix followed a few minutes later through an adjacent hole also intersecting the seam. All other holes shown in Figures 7(a) and (b) were monitored with thermocouples to establish asphalt movement towards the foundation drain.

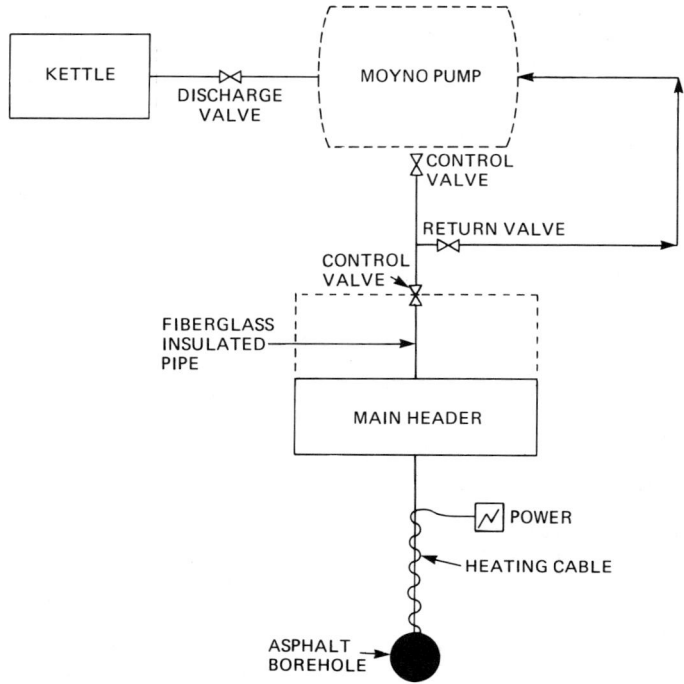

Figure 9. Asphalt Equipment Arrangement

Reduction of seepage flow was evident in the inspection tunnel within a few minutes of the start of asphalt grouting. By the end of the day, leakage from this portion of the dam had been reduced from some 3,000 IGPM (13638 lpm) to an acceptable limit, under 300 IGPM (1364 lpm). The field set up for the asphalt injection is shown in Figure 12.

In June, 1984 a similar grouting operation performed in the leakage zone beneath the northern portion of the dam reduced the leakage from about 2,000 IGPM (9092 lpm) to virtually nil.

Because of the control over the asphalt travel and its resistance to water flow the grout volumes were very modest compared to earlier work. A comparison of material quantities used during conventional cement grouting in 1982 and asphalt/cement grouting in 1983/84 is made in Table 1.

DAM FOUNDATION LEAKAGE 87

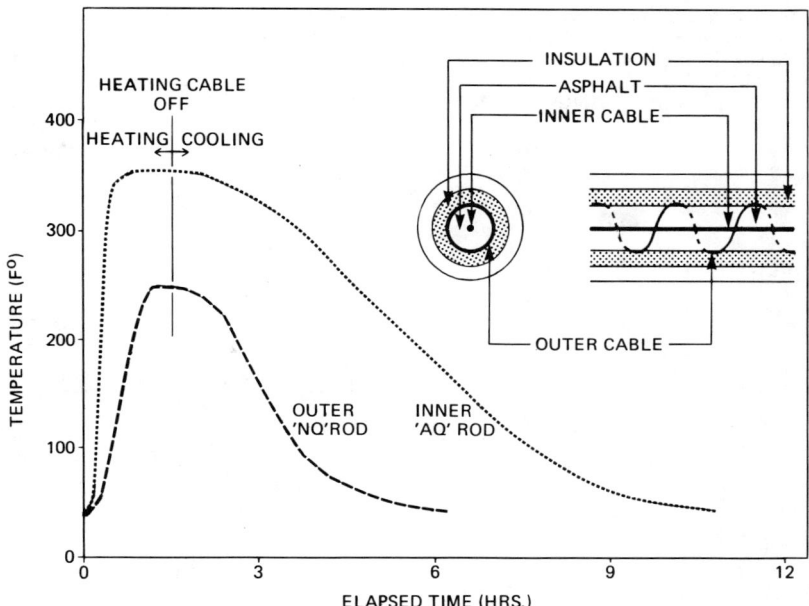

Figure 10. Heating System - Field Test

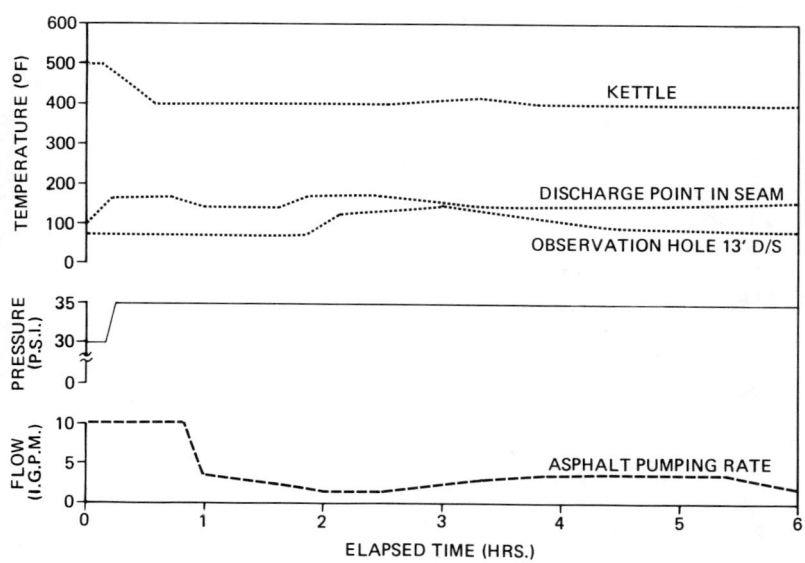

Figure 11. Pressure, Temperature, Pumping Rates - Field Results

As shown in Figure 6, no increase in leakage quantities has been observed to date (Dec. 1984). Postgrouting drilling revealed good bond between the asphalt-cement-rock, Figure 13.

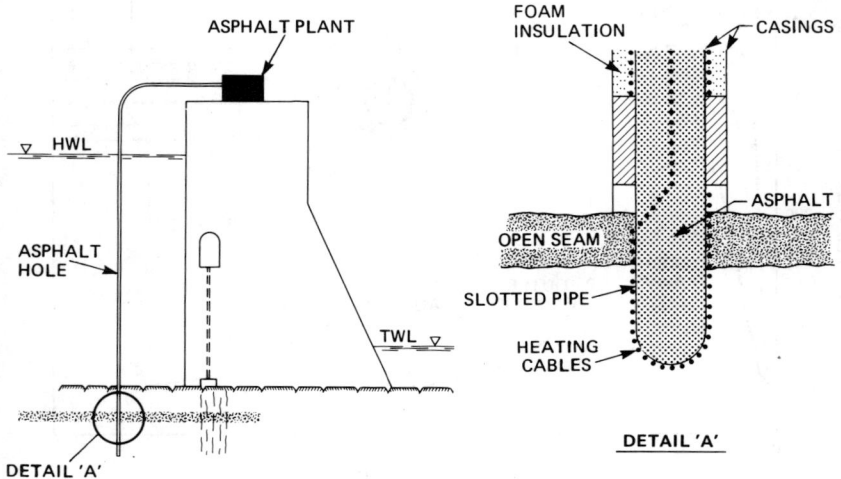

Figure 12. Asphalt Grouting - Field Setup

TABLE 1
CONVENTIONAL VS. ASPHALT GROUTING DATA

YEAR	CEMENT (BAGS)	ASPHALT (IMP. GAL.)	SAND C.F.	DURATION
1982	5 600	—	2 577	2 MONTHS
1983	280	1 584	200	1 DAY
1984	600	890	100	1 DAY

DISCUSSIONS

Based on our experience, the following information may be useful in planning future asphalt grouting operations to treat similar leakage problems:

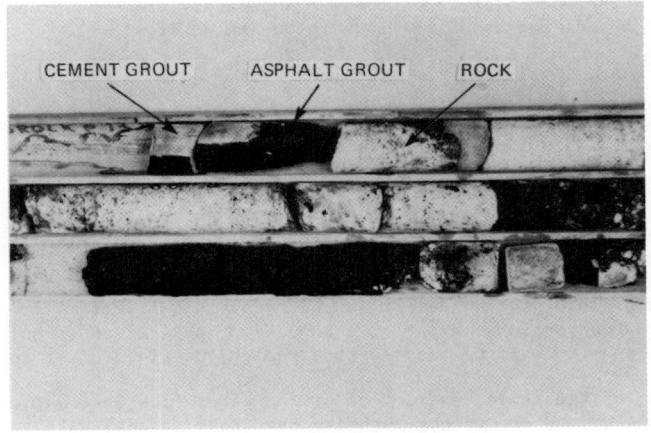

Figure 13. Postgrouting Drill Cores

Asphalt Kettle - Melting capacity required for this type of operation should be up to 10 GPM (45 lpm).

Flow Control - A proper flow meter must be designed for this type of operation.

Pump - A pump with a variable speed should be used for this type of work. To prevent overheating of the pump, a heat resistant stator and rotor, should be installed in the pump.

Valve Arrangement - To ensure safety of the operator every effort should be made to properly insulate the pipes and valves. The gauges and valves should be secured against a potential rupture.

The Heating System - The in-hole heating arrangement, such as two heating cables (inner and outer) proved adequate. As an alternative to electrical heating cables, steam was successfully used but only in the holes drilled in the concrete.

The Insert - An inflatable insert is not considered suitable for this type of operation. The insert used at Stewartville (mechanical, 60 shore hardness rubber) proved satisfactory.

Pipes - AQ-rod and NQ-casing used at Stewartville was satisfatory. However, ordinary black iron pipe ASTM120, H140 would be more economical to use. The annular space should be sufficient to allow placement of the heating cables.

Asphalt Temperature - the asphalt temperature should be based on actual field conditions encountered (i.e. large fractures - lower temperatures, small fractures - higher temperatures). If the flash point of the asphalt is exceeded, it will react violently with the water and breaks into small fragments. As shown in Figure 11 an average asphalt temperature of 400°F (204°C) was used during this operation. Thermocouples were found to accurately trace movement of asphalt in the direction of the foundation drain, Figure 11.

SUMMARY

1. The described asphalt grouting technique was successful in sealing the major leakage zone under full operating reservoir head at the Stewartville dam.

2. The asphalt grouting performed on this project proved to be more economical under the conditions encountered than conventional cement grouting alone.

3. Although field procedures for asphalt grouting differed somewhat from the conventional grouting method, its application is quite simple and it could be adapted to a variety of field conditions.

4. Because of the rapid restriction of the water flow that was made possible by the use of hot asphalt, the grout volumes were very modest compared to cement grouting. A high level of confidence can be placed in this technique with regard to predicting grout volumes and work schedule.

INDEX OF KEY WORDS

Leakage, dam, grouting, asphalt, cement, foundation drain.

ACKNOWLEDGEMENTS

The authors wish to express their appreciation to Ontario Hydro Management for their encouragement and permission to publish this paper.

Appreciation is also expressed to our many colleagues at Ontario Hydro, for their contribution to the success of this program.

Special thanks goes to Mr. Alex Nandts of ECO Geotechnical Ltd. of Saskatoon, Consultant on this project, for his contribution in developing and implementing the asphalt grouting technique.

REFERENCES

1. Deans, J. (1983). "Stewartville GS Laboratory Assessment of Grouting Materials for use in Cutting Off Ground Water Seepage", Ontario Hydro Report No. 83-458-H.

2. Deans, J. (1983). "Stewartville GS Review of Grouting Materials for Use in Cutting Groundwater Seepage", Ontario Hydro Report 83-354-H.

3. Duncan, W.M. (1972). "Stewartville Generating Station Grouting High Velocity Leakage", Canadian Electric Association, Montreal, Quebec.

4. Lukajic, B. (1983). "Stewartville GS 1983-Grouting Program, Completion Report", Ontario Hydro Report No. 84136.

5. Lukajic, B., Dupak D., Entwistle, C. (1984). "Stewartville Generating Station, 1984 Grouting Program, North End, Completion Report", Ontario Hydro Report No. 84480.

6. Smith, G. (1980). "Stewartville GS - Main Dam Leakage, Underwater Inspection and Remedial Works", Ontario Hydro Report.

CHEMICAL GROUT CURTAINS AT OX MOUNTAIN DAMS

By Edward D. Graf[1] M. ASCE, Daniel J. Rhoades[2] and Kenneth L. Faught[3]

ABSTRACT: Grout curtains were constructed in granodiorite bedrock as part of two earthfill diversion dams at a sanitary landfill site on the San Francisco, California peninsula. The curtains are three row, split spaced, and percussion drilled, utilizing a chemical grout. The resulting curtains are very effective and were accomplished at low cost within a tight construction schedule.

LOCATION AND PURPOSE OF THE PROJECT

The project is located near Half Moon Bay, California. It consists of two earth fill drainage control structures to divert surface and groundwater flows away from a sanitary landfill. Each structure consists of a low diversion dam and grout curtain in the bedrock below the dams. (Fig. 1) Structure One is located at the upstream end of current landfill operations and at the downstream end of the future landfill underdrain system. A collection system consisting of perforated pipes and relief wells was installed on the upstream side of the structure in order to collect groundwater and surface runoff. The whole collection system will be isolated from the overlying landfill by an impervious blanket. Structure Two is located immediately upstream of the maximum extent of the landfill. Surface drainage will be collected and diverted into a pipeline around the landfill at this structure. There will be a small permanent impoundment at this dam. The grout curtains constructed at both structures are intended to prevent groundwater seepage from escaping the collection system and entering the landfill material.

---

(1) President, Pressure Grout Company, South San Francisco, California

(2) Principal, Purcell, Rhoades & Associates, San Jose, California

(3) Project Engineer, Pressure Grout Company, South San Jose, California

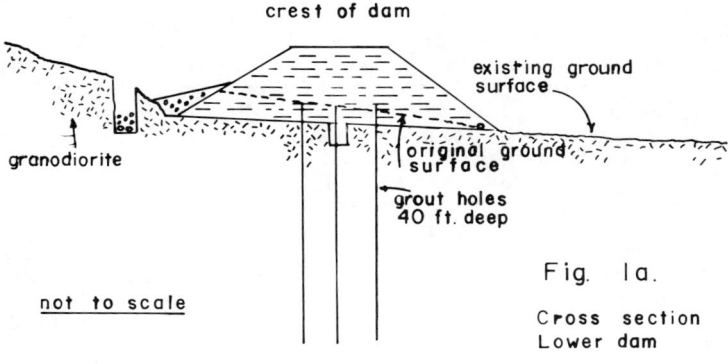

Fig. 1a. Cross section Lower dam

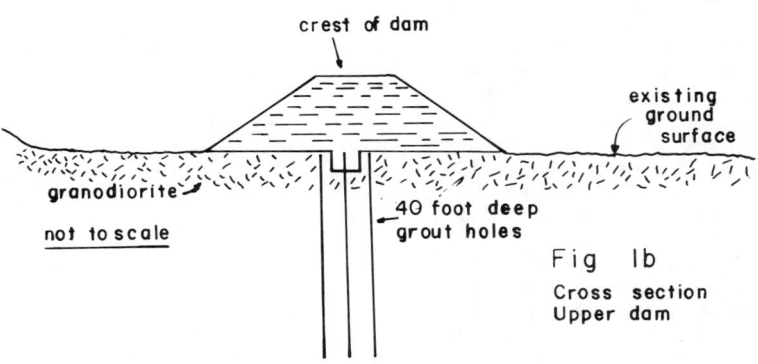

Fig 1b Cross section Upper dam

Figure 1. Cross Section of Diversion Structures

GEOLOGY OF THE PROJECT

The dam sites are in a steep walled, narrow valley with a relatively small catchment area. The valley walls are clearly marked by landslide topography of varying ages. The geology of the two sites is similar. Bedrock is a granodiorite intrusive, overlain by alluvial and colluvial deposits. The bedrock is weathered to a depth of more than 20 feet below surface. Jointing is prevalent and well developed. The near surface rock contains numerous clay filled, serpentinized, and/or slickensided joints. Open joints were found in many locations. Close jointing and microfracturing are characteristic of the bedrock as evidenced by exploration holes cited in the site selection study and environmental impact statement. The fracturing affects the engineering properties of the rock, it is easily rippable, and the permeability of the rock is controlled by fracturing.

Numerous seeps and wet areas are found along the base of the valley walls. Where the bedrock has been exposed at the sites or for landfill, these seeps were found to be coming from the bedrock. The seeps are interpreted as an indication that the dam sites are in an effluent groundwater zone.

GROUT CURTAIN DESIGN

Water that seeps into the landfill and percolates through to the leachate collection system must be treated for the lifetime of the fill (estimated to be 50 years). The surface and ground water collection system reduces the infiltration of water into the landfill and therefore reduces the cost of treating leachate. Since water that escapes from the collection system becomes expensive leachate, it is justified to expend exhaustive grouting effort to prevent seepage into the landfill. The available geologic data suggested that the bedrock is relatively tight, but that significant water movement occurs near the surface in jointed and weathered rock.

Water pressure tests conducted during siting of the landfill indicate that the upper jointed and weathered granodiorite has a permeability coefficient on the order of 5×10^{-4} cm/sec. Laboratory tests indicate a much lower permeability for the intact rock. Using a table [1] relating joint frequency, width and permeability, the observed joint frequency of between 3 and 30 joints per feet gives a range between 0.002 inches and 0.004 inches for the joint openings. This is consistent with the characteristic jointing pattern of the rock

Design criteria were established for the performance of the seepage control grouting. A curtain extending into fresh, unweathered rock to a depth of 40 ft below the base of the diversion dams and having a permeability coefficient of 1×10^{-6} cm/sec was determined to be a feasible method of reducing seepage under the structures. At this time the decision was taken to negotiate the grouting contract rather than produce detailed bid documents, and take formal bids. The grouting contractor was brought into the planning and design of the curtain at an early stage and made contributions to the planning of the curtain which helped to reduce the overall cost of the project while increasing its effectiveness.

Choice of Grouting Materials

The material used for grouting was selected based on consideration of the following criteria:

A. In order to minimize seepage through the curtain, the grout should be able to permeate very fine fractures.

B. The grout curtain should remain impermeable over the service life of the project, and the grout material should be insensitive to leaching.

C. The radius of travel of the grout material should be controllable to within a few feet.

D. The material should exhibit full volume set without bleed, shrinkage or syneresis.

E. The grout should develop adequate strength to withstand the potential full groundwater head after construction of the landfill.

F. The grout material should be economical.

Portland cement based grouts are traditionally used for dam grouting, but it is recognized that they do not permeate fine fractures, are dissolved by acids, and are subject to bleeding/shrinkage in normal practice. The overall reduction of permeability of rock mass grouted with cement base grout cannot drop below the level of the secondary permeability which arises from fractures too fine to be permeated by cement particles. A joint opening of 0.01 inches is often cited as a lower limit for permeation by cement particles.

Silicate based grouts, without particles in suspension, exhibit superior permeation characteristics than cement grouts but are subject to leaching and long term syneresis. Under the influence of strong alkalis, most silicate systems are subject to dissolution. The long term effect of leaching and/or syneresis on the curtain could cause a serious increase in permeability.

Organic resin grouts have excellent permeation characteristics, radius of permeation control through control of gel times, full volume material set, resistance to chemical leaching attack, variable strengths, and they do not exhibit syneresis. The cost varies considerably for various resin types.

Acrylamide and acrylate base grouts have very low viscosity which enable them to penetrate the very fine fissures, and they have excellent chemical stability. They have excellent gel time control over a wide range, but in the end were not considered because of their relatively high material costs which would have increased the cost of the grout curtain.

The system selected for this project is marketed under the name R.E.G. It has favorable permeation characteristics, good leaching resistance, good strength although brittle, full volume set, controllable set times, lack of syneresis, and low cost. In the dilution used, the material can permeate very fine fractures and/or slightly silty sand. Permeability tests on sands grouted with similar products indicate that K values of 10^{-7} cm/sec, or less, are possible. Data on the service life of the material are sparse because it has not been in general use for more than about twenty years. Review of its chemical properties indicates that an indefinite service life greater than the project life should be expected. The R.E.G. grout is economical. When mixed at the dilution used for this project, it will produce a gel that can be competitive to an equivalent volume of cement grout.

Choice of Grouting Pattern

Experience with grouting similar rock indicated that the required hole spacing might range from two feet to more than six feet. Therefore a three row pattern of holes spaced four feet each way was established as a preliminary design pattern. The holes were to be drilled and grouted by the split spacing method starting with a primary spacing of sixteen feet. It was clearly understood that the four foot spacing could be changed based on the observed performance of the primary and secondary holes.

Stage length were defined as a maximum of 20 feet, or at drill water loss or gain. This was done to reduce the risk of plugging fractures with drill cuttings.

Choice of Drilling Method

The cost of drilling is frequently the most expensive part of constructing a grout curtain. In this case, where four foot spacing of the holes was expected, the drilling was anticipated to be more than one half of the curtain cost. Accordingly, the choice of the drilling method was strongly influenced by the cost.

Percussion drills modified to use water as a drilling fluid were selected. This produces a cleaner hole than air drilling. It also allows the observation of fluid loss (or gain) which was to be used to indicate whether a stage should be grouted at a particular depth.

Choice of Grouting Pressure

The grouting pressure was established to promote the maximum permeation of the rock in the minimum pumping time without causing fracturing and uplift of the foundation. The pressure was limited to a maximum of 100 psi. This pressure will have the effect of dilating fractures, and will cause uplift of slabby rock near the surface. Dilation of fractures has the effect of temporarily increasing the permeability of the rock, and promotes grout spread. As the pressure is released, elastic rebound closes the fractures and helps to keep the grout fillings intact. Uplift of slabby rock allows grout to fill the joints between the rock slabs, but is harmful to the extent that contact and friction between slabs may be reduced.

In order to minimize the risk of damaging effects, fast gel times and controlled pumping rate is used. The pumps are operated at a speed which will not allow the pressure to reach the maximum until late in the injection. The onset of gellation dissipates the pressure within a few feet of the hole, and the total area along a potential slab that is subjected to full grouting pressure is not likely to be large enough to cause damage. Good judgement and experienced personnel are required to optimize this technique.

The grouting pressure that was used is a significant departure from pressures generally specified on American dam grouting projects. Grouting pressure is often specified in ranges from the conservative "not to exceed the weight of overburden" up to two times the weight of overburden in competent rock.[2,3] The rationale for using the high injection pressure is that the anticipated low permeability of the rock would cause impractical injection rates at lower pressure.

Choice of Gel Time

Manipulation of gel time is a technique used to control the radius of grout injection. The permeability of the rock will limit the maximum grout injection rate, but the radial distance the grout will travel is a function of viscosity of the grout. If a low viscosity grout is pumped into the hole, the radial flow will continue until the grout begins to gel. As the gel time nears, the viscosity of the grout rises and there is a large head loss. Use of a short gel time will limit the distance the grout can travel and causes large head loss within a short distance from the hole.

Initially, the gel time was set at 2-3 minutes based on the contractor's experience and judgement. During the course of construction, observations of the grouting pressure and the distance of grout travel from the injection point were used to evaluate if shorter or longer gel time was needed. The appearance of grout at the surface or in nearby grout holes is the usual means of observation. No changes in the gel time were made during the construction of the project.

CURTAIN CONSTRUCTION

The curtains were constructed during the summer of 1984. The contractor was retained to perform the work on a time and materials contract. The engineer provided continuous inspection during the construction period. Payment for the work was on a completed contract. The lower dam was completed by one crew working five ten hour shifts per week. The upper dam was completed by working two twelve hour shifts daily.

The total grout takes are shown in the following table:

Table 1

Ox Mountain Project, Summary of grouting

Location	Footage	Stages	Grout	Take
Lower Dam	1900 lf	88	144.7 ft3	0.08 ft^3/ft
Upper Dam	2170 lf	106	312.0 ft3	0.14 ft^3/ft

The curtains are three rows each with staggered holes spaced four feet each way and 40 feet in depth. (Fig. 2) The lower dam (Fig 1) is 68 feet wide with a single row of holes fanned into the abutments. (note: Row 3, the center row, was partially deleted because of low primary takes.) The upper dam (Fig 1) is 52 feet wide with a fan of holes into each abutment on the upstream and downstream rows.

The work was completed for a total of $94,230 for an overall cost of $12.15 per vertical square foot of curtain. The use of longer shifts and around the clock operation at the upper dam resulted in some efficiencies and a small savings in construction costs.

The grout takes of 0.08 and 0.14 ft^3/ft are low. They are consistent with the geologic conditions, and the tight fracturing of the rock. Grout take is one of the methods used to measure the effectiveness of the grouting operation. In material where there is a good interconnection of joints

CHEMICAL GROUT CURTAINS

Grout Hole Locations Lower Dam

Angle Holes			Angle Holes		
39P	54 gal	22.5°	42P	24 gal	22.5°
40T	50 gal	45°	43T	17 gal	45°
41T	30 gal	67.5°	49T	17 gal	67.5°

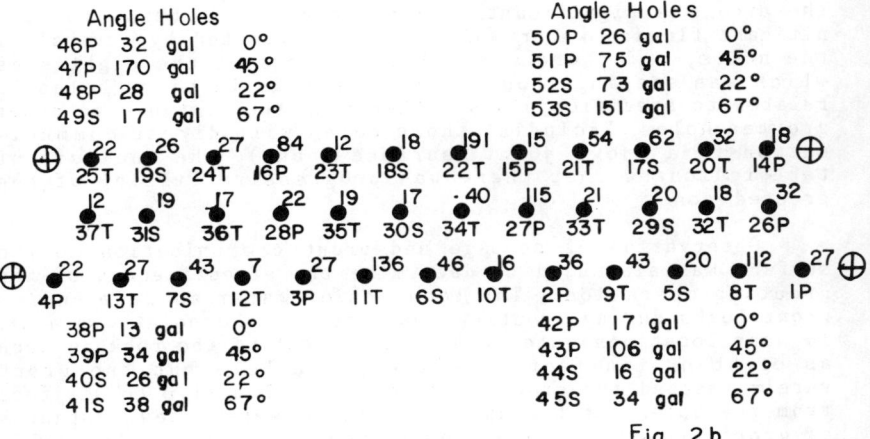

Legend
- ●40 — Grout consumption in gallons / Grout hole location
- 2P — Grout hole number
- ⊕ — Angle hole location

Fig. 2a.

Grout Hole Locations Upper Dam

Angle Holes			Angle Holes		
46P	32 gal	0°	50P	26 gal	0°
47P	170 gal	45°	51P	75 gal	45°
48P	28 gal	22°	52S	73 gal	22°
49S	17 gal	67°	53S	151 gal	67°

38P	13 gal	0°	42P	17 gal	0°
39P	34 gal	45°	43P	106 gal	45°
40S	26 gal	22°	44S	16 gal	22°
41S	38 gal	67°	45S	34 gal	67°

Fig. 2b.

Figure 2. Grout hole locations and grout consumptions.

and voids, the grout will spread in a roughly cylindrical zone around the injection holes. As the distance from a hole to the nearest grouted hole is reduced to a point where the zones overlap, there will be a marked reduction in grout take. In materials where the joints and voids are not well interconnected, the reduction in grout take is not as clearly defined as the hole spacing is decreased. The rock at both sites represents the latter case.

The average take in the center row holes was reduced from the take in the first two rows, but the scatter between primary holes and the later holes masks any trend toward lower takes. The observations during the grouting indicated that there were only a few joints which would take large quantities of grout. These areas were identified by the location of grout return along a few well defined joints and communication to previously drilled holes. On the left side of the lower dam between holes 7 and 10 and between 24 and 27 there was an interconnection with a group of joints. Hole 38, which was the check hole in that area, did not show evidence of interconnection and the area was determined to be completed. On the right side of the curtain there was no clear evidence of connection to any group of joints but grout did move from the tertiary holes to adjacent holes. Holes 35, 36 and 37 did not show any evidence of communication to previously grouted holes in the first two rows, so this area was also determined to be completed.

The takes at the upper dam were larger, and more nearly resembled the first case, but the effect of a few joints in the area was significant. There were two joints with significant flowing water which were intercepted by several of the holes, and there was a diffuse area on the right side which was seeping about 1 gpm. The take in these holes is related to specific joints rather than the distance to other grouted holes. Excluding those holes with direct communication to a known joint (surface leaks), the analysis of take indicates that there was progressive filling of the grouted zone.

Observation of seepage and grout communication to the surface was also used to determine the effectiveness of the grouting operation. The location of water seepage and/or grout leaks during grouting was noted by the point operator. In many locations water was squeezed out of the rock as much as eight or ten feet from the grout hole but the grout rarely reached the ground surface more than five to six feet from the hole. At the upper dam there were a few instances of grout traveling more than 10 feet along a single, well-defined joint.

OBSERVATIONS

During construction of the curtain, there were indications that the grouting operation was proceeding as planned and that the work was having an effect on the groundwater movement. The most noticeable effect was sealing of joints that were discharging water prior to grouting. The lower dam was not instrumented to observe piezometric levels because of the great depth of burial under landfill but observation wells are planned for installation at the upper dam as part of a future contract.

The following observations and interpretations were made by the grouting engineer:

1. Grout communication to the surface was along well defined joints. These joints usually were intercepted by more than one hole.

2. Water was squeezed out of joints in the rock, and completed grout holes at distances up to 8ft. from the injection point. Grout did not always reach the points where water was discharged during grouting.

3. There were no observed instances of uplift or heave of the foundation as a result of the grouting pressure.

4. As the grouting proceeded, seeps in the downstream area reduced flow or disappeared.

5. As the grouting proceeded, seeps in the upstream area increased in flow and new seeps appeared. (photo 1)

6. Isolated joints that were flowing water were intercepted, and grouted to complete dry condition.

7. During excavation of the keyway for both dams, grout was found filling nearly all open joints, and in very fine fractures throughout the rock. (Photo 2)

8. The keyway excavation in both areas was dry. This is noteworthy because there were significant seeps in both areas prior to grouting.

Photo 1.
Seeps forming upstream of the grout curtain.

Photo 2.
Grout filling joints in the keyway excavation.

CONCLUSIONS

The project was successful in meeting the design goal of limiting seepage below the water collection system, and entering the sanitary landfill The construction was completed in a short time, and at a very reasonable cost. The following are the major conclusions drawn:

1. The use of chemical grout is feasible for sealing tightly fractured rock.

2. Chemical grout can be an economic alternative to traditional cement base grouts for dams, and can be more effective.

3. The high grouting pressure did not have adverse effects on the foundation rock.

4. Percussion drilling, wethead, did not seal the fine fractures from grout permeation.

5. The observed performance of the grout curtain is satisfactory.

REFERENCES:

1. E. Hoek and J.W. Bray, "Rock Slope Engineering" The Institution of Mining and Metallurgy, London, 1974.

2. Houlsby, A.C., "Cement Grouting for Dams" Grouting in Geotechnical Engineering, ASCE, Feb. 1982

3. Hilf, J.W., and Deere, D.U., "Summary-Panel discussion of Dam Foundations", Rock Engineering for Foundations and Slopes", Proceedings of Specialty Conference of Geotechnical Engineering Division, ASCE, 1976

EMBANKMENT FOUNDATION DENSIFICATION BY COMPACTION GROUTING

by

Wallace Hayward Baker[1] Member ASCE

ABSTRACT

Compaction grouting is an attractive alternate for in-place densification of liquefiable soils below existing structures. Compaction grouting application concepts and procedures are presented for the important case of embankment dams founded on liquefiable sandy soils. Preliminary results of two full-scale test programs are presented to illustrate the proposed approach.

PROBLEM BACKGROUND

One component of modern embankment design is the assurance that foundation materials will not liquefy during earthquake activity of reasonable magnitude. This requires an evaluation of foundation support materials beyond simple static stability, and was not a part of embankment design until the last few decades.

The National Dam Safety Inspection Program has identified several decades-old embankment type hydraulic dams whose support soils pose stability problems. An important number of these potential stability problems involve possible liquefication of loose sandy materials under earthquake conditions. Among the options available to remedy this situation in existing dams are: (1) construction of a new adjacent embankment on acceptable foundation materials (embankment replacement); (2) in-place densification of the problem liquefiable soils below the dam (foundation rehabilitation); or (3) combinations of the above.

---

[1] President and Chief Engineer, GKN Hayward Baker Inc., Odenton, Maryland.

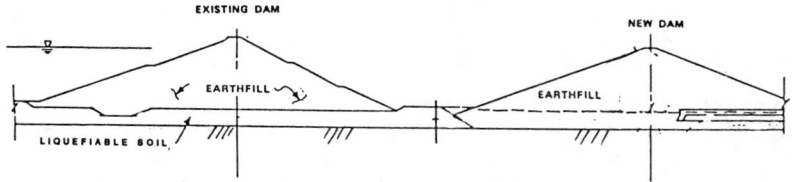

(a) NEW ADJACENT EMBANKMENT DAM

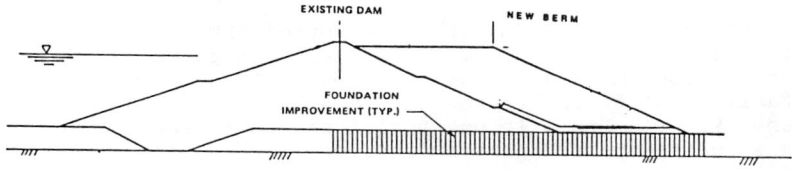

(b) IN-PLACE DENSIFICATION OF DOWNSTREAM FOUNDATION SOILS

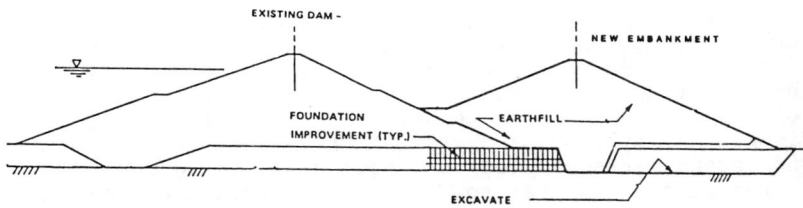

(c) COMBINED NEW EMBANKMENT AND IN-PLACE DENSIFICATION

FIGURE 1. Alternative Solutions for Existing Embankment Dam on Potentially Unstable Foundation.

These three alternative approaches are depicted in Figure 1, showing the cross-section of a typical embankment dam where a layer of liquefiable sands is located in the foundation. Figure 1(a) shows the construction of a new dam, where the problem soil layer has been excavated and replaced with compacted material. Figure 1(b) shows a rehabilitation program where the problem soils have been treated in-place, generally down slope from the core. A berm has been added to the embankment since it is judged that the problem soils on the upstream side of the embankment would be difficult and costly to treat. Figure 1(c) shows a combination of the two solutions, with some excavation and replacement and some in-place densification. The actual design chosen would be based on the relative costs of new embankment construction and the in-place treatment procedure.

Candidate methods for in-place densification of sandy soils include Vibro-Compaction, Dynamic Compaction, Compaction Piles, Chemical Grouting, and Compaction Grouting. Based on cost, safety, and effectiveness, Compaction Grouting is one of the most attractive alternatives for densifying liquefiable soils below existing embankments.

Compaction grouting densification of embankment support materials was proposed by the writer in a presentation to the "Symposium on Remedial Measures to Improve the Dynamic Stability of Earth Structures," at the ASCE Annual Convention in New Orleans, Louisiana, in September 1982. This current paper further describes the concept and performance guidelines of compaction grouting for embankment foundation densification and presents some preliminary results of two full-scale test programs now in progress.

The most extensive of the two test programs is now being carried out at the West Pinopolis Dam in Monks Corner, South Carolina, by Shannon and Wilson, Inc., and GKN Hayward Baker Inc. under the direction of the Savannah District of the Corp of Engineers. The geotechnical situation shown in Figure 1 is typical of a portion of the West Pinopolis Dam. The second, more limited test program is being carried out at the Steel Creek Dam now under construction at the Savannah River Plant, Jackson, South Carolina. This test program work is being done by GKN Hayward Baker Inc. under subcontract to J. A. Jones Construction Company, and is being directed by the Corps of Engineers, aided by Geotechnical Engineering, Inc. Full details of these two test programs will likely be reported following completion.

COMPACTION GROUTING CONCEPT

Compaction Grouting is the injection under pressure of a stiff mortar grout to displace and compact adjacent soils. Figure 2 depicts a compaction grout bulb with the displaced and compacted soil surrounding it. In contrast to permeation (chemical) grouting, the influence of the grout extends well beyond the grout mass, involving soil volumes up to 20 times the grout volume.

The procedure has been used for remedial grouting to solve footing and floor slab settlement problems for over 30 years (Warner, 1982), and for settlement control for soft ground tunneling for the past ten years (Baker et al, 1983). Only recently has Compaction Grouting been applied to site improvement problems (Schmertmann and Baker, 1983; Graf, 1984; Donovan, 1984).

Compaction Grouting for embankment foundation densification differs greatly in scope from classical Compaction Grouting for remedial structural foundation projects. Embankment foundation densification treatment depths are usually significantly greater, the grout volumes are much larger, and the injection rates can be up to 10 times faster.

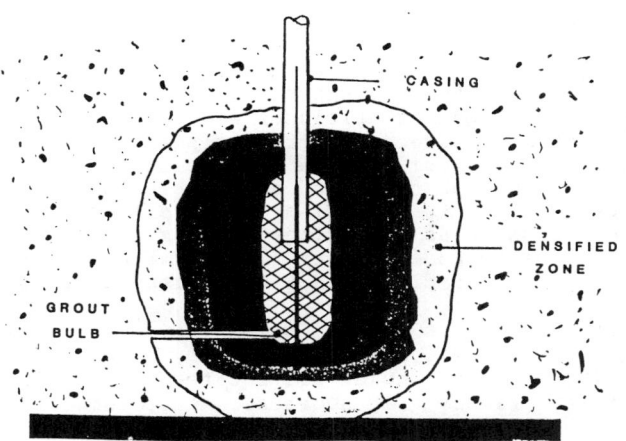

FIGURE 2. Compaction Grout and Adjacent Densification Zone.

A compaction grout mix typically consists of a harsh mixture of silty sand, cement and enough water to result in a cone slump of 1 to 3 inches (25 to 150 mm). Actual compaction grout design mixes have also included one or more of the following additional ingredients: fly ash, gravel, lime rock, bentonite, and water reducing agents. But in each case the resulting mortar has been a stiff, dense, low plasticity mix that will exhibit water loss under pressure to produce a relatively rapid stiffening of the grout in place and result in a controlled grout bulb rather than an uncontrolled lens.

Grout pipes are typically 2 to 4 inches (50 to 100 mm) in diameter and are tightly sealed into the ground to prevent grout leaks and pipe extrusion during grout injection under relatively high pressures of 150 psi up to 1000 psi (1 to 7 MPa). An initial injection cavity is typically made by extending the grout hole some 8 inches to 10 feet (20 cm to 3 meters) beyond the end of the grout pipe or by raising the grout pipe. Compaction grout injected under pressure into the cavity at the end of the grout pipe laterally displaces and compacts the adjacent ground until either heave occurs or the pressure rises enough to cause stiffening (sand blocking) of the grout mass in place. In the latter case, sudden apparent injection pressure increases occur. The

next injection location is treated either by drilling
through the previously placed grout mass to extend the grout
hole to the next lower zone (top-down method), or by
withdrawing the grout pipe in increments and injecting in
the space left below the raised casing (bottom-up method).

INJECTION PLANNING

Grout Hole Layout. A compaction grouting densification
program is carried out in a planned grid pattern of grout
pipes, injected in phases at closer and closer spacings.
The ideal distance between the grout pipes depends on the
soil conditions, maximum acceptable pumping pressure
(related to overburden depth and acceptable heave), and
required density. Actual final grout pipe spacings have
varied from as close as 4 x 4 feet (1.25 x 1.25 m) square
grids to as wide as 25 x 25 feet (1.8 x 2.5 m) square grids.
Typical final grid spacings are 6 to 10 feet (1.8 to 3 m).

Initial grout holes are termed "primary holes" and are done
on wide grid spacings. After all primary holes are grouted
in an area, "secondary" holes are located and grouted at the
centers of the primary grid squares. After completion of
primary and secondary holes, "tertiary" holes are located
and grouted at the centers of the resulting primary-
secondary squares. After tertiary holes, the final grid
size is one-half the width of the initial primary hole grid
spacing. Figure 3 shows a typical layout plan.

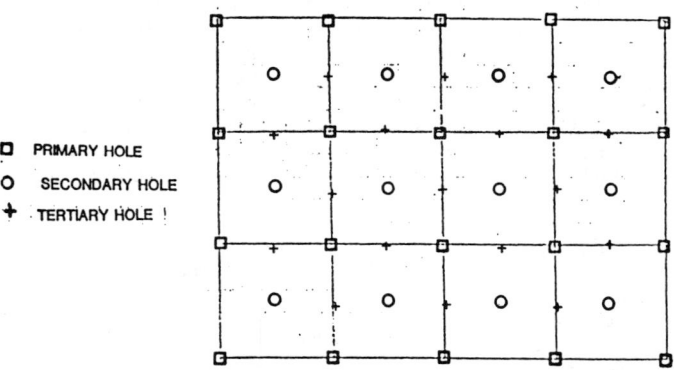

FIGURE 3. Primary, Secondary and Tertiary Hole Layout Plan.

Note that over a large square grid area, the number of
secondary grout holes is equal to the number of primary
grout holes. Also, the total number of tertiary holes
equals the total number of primary and secondary holes.

Grout Quantities. An initial estimate of the volume of required grout is made based on the anticipated necessary change in density within the target treatment zone. The required grout quantities can be as low as a few percent to as high as 20 percent or more of the total target zone.

Determination of in-place density and approximate maximum density for the target treatment soils will give an indication of the maximum practical net displacement volume that can be expected from the compaction grouting efforts. Maximum relative density is only useful for very clean sands. For example, at the Pinopolis test site, the loose sands contained only about 5 percent of fines passing the No. 200 sieve, but this fine fraction had a Plasticity Index of 200-300. For this unusual material, the maximum density as determined on a shake table was only about 7 percent higher than the in-place density, whereas the maximum Proctor density was about 15 percent greater. Potential maximum grout takes should be based on the higher density found in the Proctor compaction test, as demonstrated by the Pinopolis Dam test results.

In classical foundation rehabilitation compaction grouting where the goal is to stop on-going footing or floor slab settlements, grouting at a particular location usually continues until pressure refusal or slight surface heave is noted, regardless of the grout quantities involved. In foundation densification work, the probable total grout quantities should be predicted and the target grout volumes should be established and assigned in a rational way to the primary, secondary and tertiary grout holes.

Larger target quantities are usually specified for the primary and secondary holes, with reduced quantities anticipated for the tertiary holes. The primary and secondary grout holes might be limited to 1/3 of the total anticipated grout take each, and the tertiary holes limited to not more than 1/4 of the total target grout take each. Such a target grout quantity distribution is shown below.

Hole Type	Grout Holes	Target Grout Take
Primary	25%	1 x 33% = 33%
Secondary	25%	1 x 33% = 33%
Tertiary	50%	2 x 25% = 50%
All Holes	100%	116%

If all holes in the above example take the maximum target values, the total results in a maximum take of 116 percent. Actual grout takes may vary considerably from the target grout takes.

Injection Termination Criteria. In practice, the target grout take criteria are usually coordinated with maximum pressure and heave criteria. Maximum injection pressure criteria are usually initially established using past experience rules-of-thumb, and then adjusted during the early grouting episodes at a specific site.

Above a particular pressure range, a specific mix design will exhibit sudden water extrusion with associated stiffening (sand blocking). Maximum injection pressures should be set at least 100 psi (about 0.70 MPa) below this apparent sand blocking pressure. Where sand blocking is occurring before adequate ground densification, the mix design must be adjusted to obtain a higher water extrusion (sand blocking) pressure.

If injection pressure is plotted versus the log of cumulative grout volume, (or log of time assuming constant rate of injection), most granular materials will show a steady pressure increase along an approximate straight line. When the pressure versus log-time curve suddenly steepens, sand blocking has usually occurred. When the grout pressure suddenly stops increasing, heave has commenced or a void has been penetrated. In the latter case, grouting should be stopped and the cause for pressure loss investigated. The above pressure versus log-time trends are shown conceptually in Figure 4.

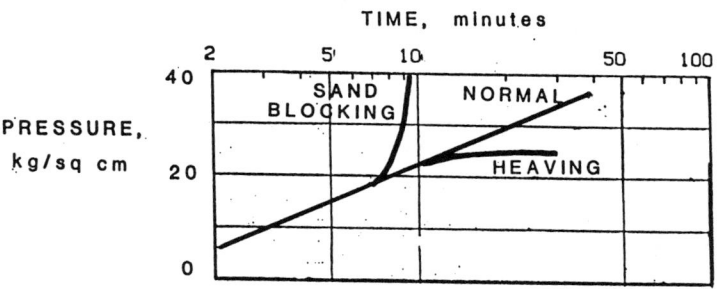

FIGURE 4. Conceptual Pressure Versus Log-Time Curves.

When significant surface heave results, the rate of lateral compaction is greatly reduced and pressure increase stops. Both incremental heave limits (over a given injection stage depth) and cumulative surface heave criteria are usually established. When grouting on a sloping embankment, lateral displacement can easily occur and may limit effectiveness.

The following injection termination criteria were used for Steel Creek Test Plot No. 3.

"Compaction grouting shall continue at each 5-foot hole stage until one of the following criteria occurs:

(a) Target grout volume for the stage has been met.

(b) Maximum surge pressure of 400 psi (2.75 MPa) or greater occurs consistently at the current stage.

(c) Maximum backpressure of 200 psi (1.38 MPa) or greater occurs at the current stage.

(d) Surface heave caused by the current grouting stage exceeds 1/4" (6.4 mm) when cumulative heave of a point located within 5' (1.5 m) of the grout hole is less than 3/4" (19 mm). "Cumulative heave" is only that for the current grout hole.

(e) Surface heave caused by the current grouting stage exceeds 1/8" (3.2 mm) when cumulative heave of a point located within 5' (1.5 m) of the grout hole is equal to or greater than 3/4" (19 mm). "Cumulative heave" is only that for the current grout hole.

(f) Lateral down-slope movement at the ground surface exceeds 1/8" (3.2 mm) for current grout stage.

(g) If cumulative surface heave at any point exceeds 3" (76 mm), Engineer will decide if grouting continues.

(h) If cumulative lateral movement at the ground surface exceeds 1/2" (12.7 mm), Engineer will decide if grouting continues.

The above injection criteria are project specific and given as an example only.

GROUT TAKE EVALUATION

Grout take distribution can be usefully evaluated by considering single primary-secondary grids with tertiary holes at the centers of the grids. The injected grout quantity at a given stage for each primary or secondary hole is assigned proportionally to each adjacent grid square (25 percent to each grid in the case of a square grid). The full 100 percent of the tertiary hole grout quantity is also assigned to its grid square.

This method of proportional grout take assignment to evaluation grids is shown in Figure 5. This method provides a rational basis for comparing grout takes from the primary-secondary injection phase with the grout takes for the tertiary holes. Typically, tertiary grout takes should be somewhat less than the combined primary-secondary grout takes.

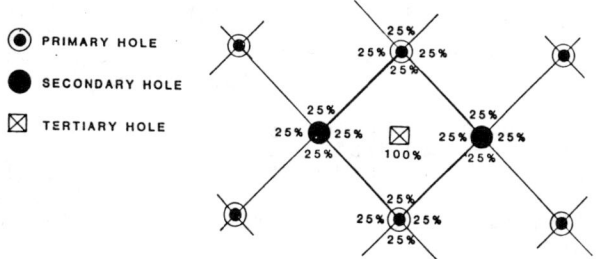

FIGURE 5. Grout Volume Assignments for Grid Analysis.

Using the area of the grid being evaluated and the associated depth of treatment, the affected soil volume can be calculated and compared to the grid grout take to obtain the grid grout percentage for that grid square.

EMBANKMENT FOUNDATION DENSIFICATION

A sample grout take distribution plan is shown in Figure 6 with data taken from the central grid squares of Test Plot No. 3 at the Pinopolis Dam test site. The average grid grout take percentage for the highlighted four adjacent grids is 15.2, and the average percentage take for each grid is consistent. The tertiary grout takes are substantially less than the primary-secondary combined grout takes, in the ratio of about 1:3.

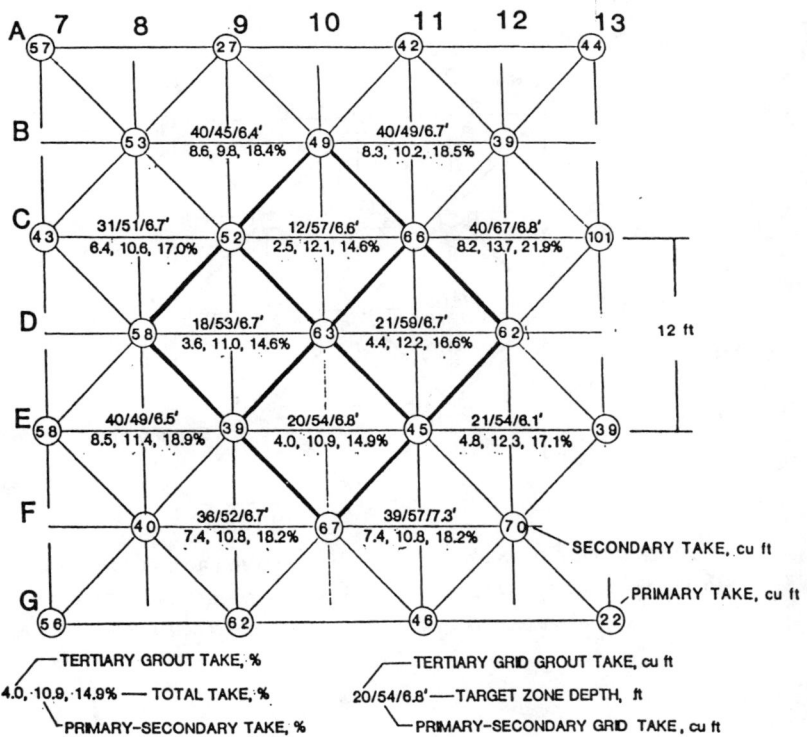

FIGURE 6. Grout Take Grids for Pinopolis Test Plot No. 3.

Shown in Figure 7 is the grout take distribution plan for the 40 to 45 foot (12.3 to 13.8 m) depth stage at the Steel Creek Dam test Plot No. 3. Inconsistency in injected volumes between adjacent holes (compare hole 1-C at 8 cubic feet with hole 2-B at 44 cubic feet) suggests premature sand blocking refusal in some holes. The average grout take within the four adjacent evaluation grids for this stage was only 4.7 percent, compared to a target grout take of 7.5 percent. The relatively low changes in penetrometer resistance after grouting, shown later in Figure 14, reflect this reduced grout take.

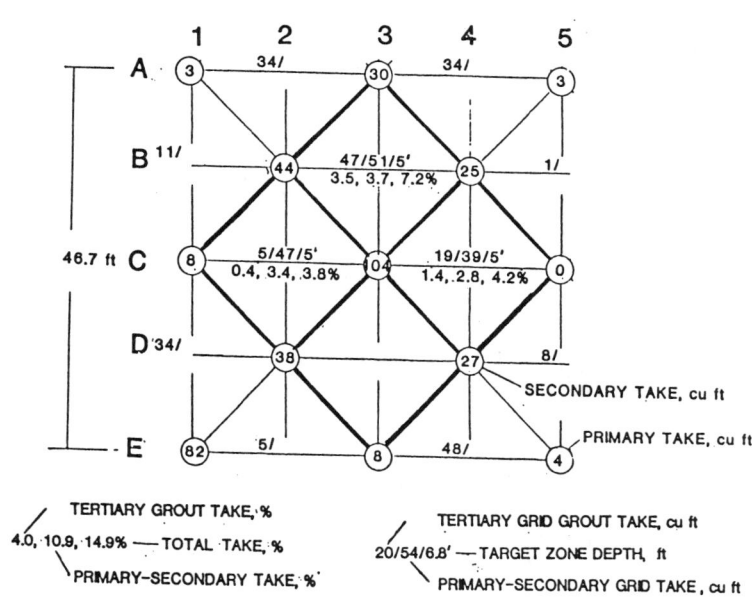

FIGURE 7. Grout Take Distribution Grids at Steel Creek Test Plot No. 3, for the 40' to 45' Stage.

PINOPOLIS DAM TEST RESULTS

The compaction grouting test program at the Pinopolis Dam is being carried out from a specially constructed berm just downstream of the existing dam. A cross-section of the berm is shown in Figure 8. The berm simulates conditions below the dam and provides necessary overburden confinement to permit high grout pressures with little heave, but without the need to work directly below an active dam before the method is successfully demonstrated.

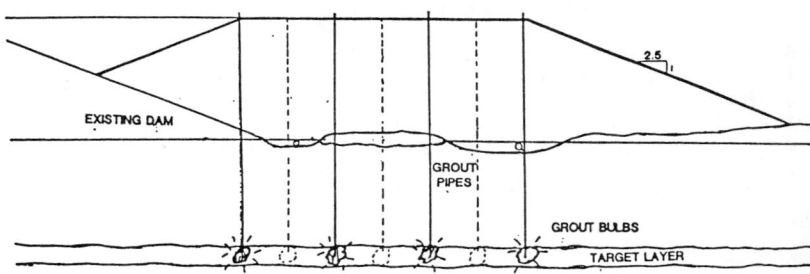

FIGURE 8. Cross-Section of Pinopolis Dam Test Plot Berm.

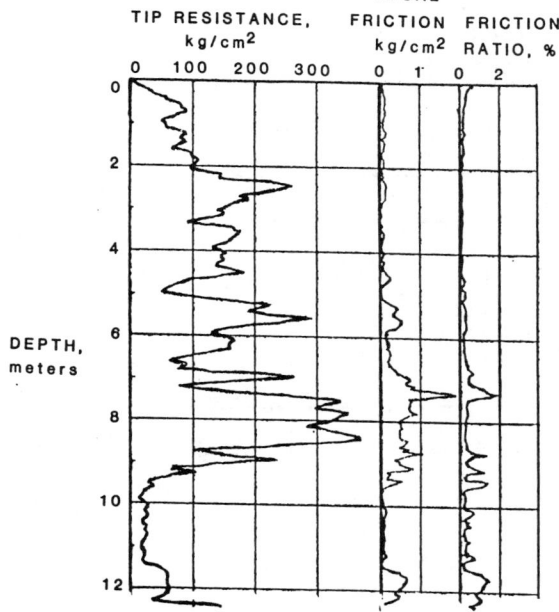

FIGURE 9. Electric CPT Profile at Pinopolis Dam Test Berm.

Figure 9 shows electric cone penetrometer test (CPT) results through the temporary embankment. The target treatment zone of about 2 meters of loose sand varies between 9 to 12 meters in depth at this location, between an overlying very dense sand layer and underlying medium stiff marl, and is representative of the problem soils under the dam.

In Figures 10 through 13 the in-situ penetrometer, standard penetration and dilatometer test results for the target treatment zone are shown in expanded scale.

Before-and after-grouting electric cone penetrometer test tip resistance results are shown in Figure 10. Before grouting tip resistance values of about 20 kg/sq cm increased to about 85 kg/sq cm after primary-secondary grouting with about 11.6 percent grout displacement. Tip resistance further increased to about 125 kg/sq cm after tertiary grouting with an additional 3.6 percent grout.

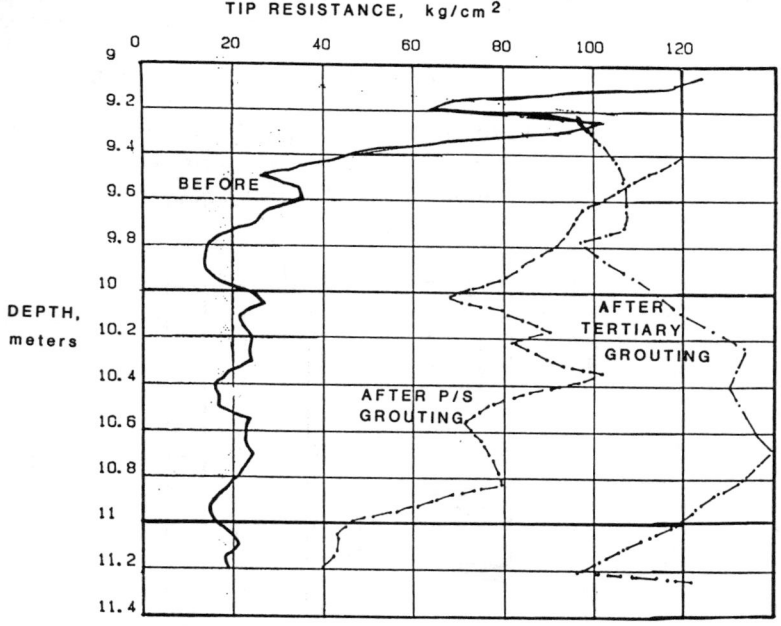

FIGURE 10. Representative Penetrometer Tip Resistance Values for Pinopolis Dam Test Target Zone.

Before-and after-grouting Standard Penetration N values are shown in Figure 11. Uncorrected N values increased from about 4 to about 25 after primary-secondary grouting. Post tertiary grouting N values are not yet available.

The very substantial relative increases in cone penetrometer tip resistance values and Standard Penetration resistance (N) values due to compaction grouting suggest corresponding increases in resistance to liquefaction under cyclic (earthquake) loading. But these changes in penetrometer and Standard Penetration test values correspond mostly to soil density increases and do not adequately reflect increased resistance to liquefaction due to large increases in lateral stresses caused by the compaction grouting.

The flatplate dilatometer test (DMT) measures local bearing capacity during advancement of the blade and the stresses required to initiate lateral movement of the flexible steel membrane and displace it about 1 mm. Interpretation of the test results involves the bearing capacity and three index parameters: the Material Index, I_D; Horizontal Stress Index, K_D; and the Dilatometer Modulus Index, E_D.

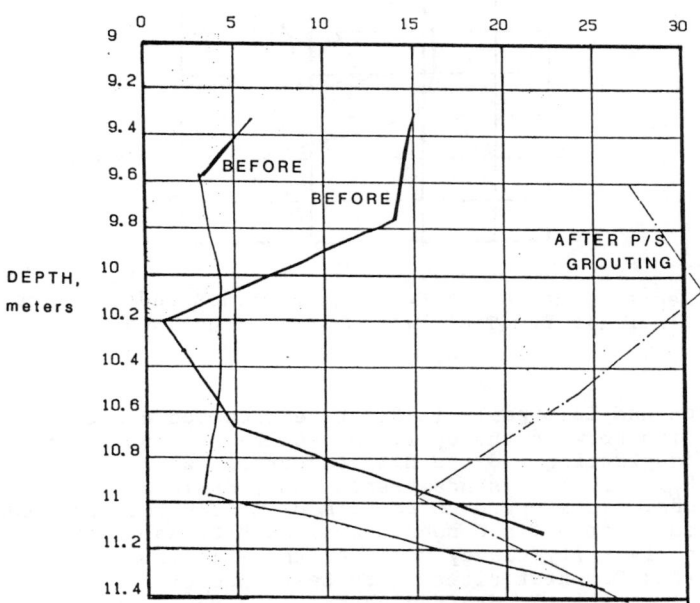

FIGURE 11. Representative Standard Penetration Test Results for Pinopolis Dam Test Plot No. 3 Target Zone.

K_D is defined as the value of the corrected initial membrane lift-off pressure less the static pore pressure, divided by the effective overburden pressure. The before- and after-grouting DMT Horizontal Stress Index (K_D) values are shown in Figure 12 in semi-log scale.

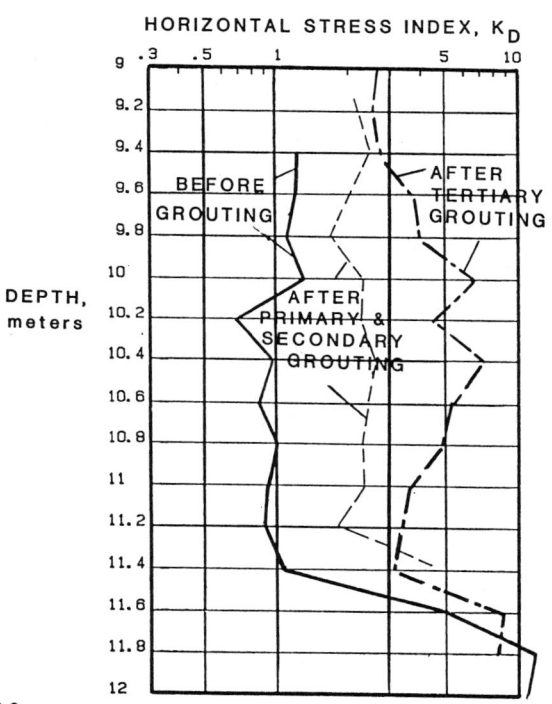

FIGURE 12.

Representative DMT Horizontal Stress Index Values for Pinopolis Dam Test Plot No. 3 Target Zone.

Robertson and Campanella (1984) have proposed a preliminary correlation between the cyclic stress ratio required to cause liquefaction and the dilatometer Horizontal Stress Index, K_D, for clean sands tested under drained conditions. This approach appears promising for DMT tests performed under drained conditions, but its application to the slightly silty/clayey sands at the Pinopolis Dam and Steel Creek Dam test sites where test conditions may be under only partially drained needs to be confirmed.

DMT constrained modulus values, M, are calculated from E_D and K_D, and are used to calculate settlements. In Figure 13, before-and after-grouting DMT constrained moduli values are shown on semi-log scale. Before-grouting constrained moduli values of between 20 to 100 kg/sq cm increased to an average of nearly 500 kg/sq cm after primary-secondary grouting, and to a maximum of about 2,000 kg/sq cm after tertiary grouting, giving a maximum 70-fold increase.

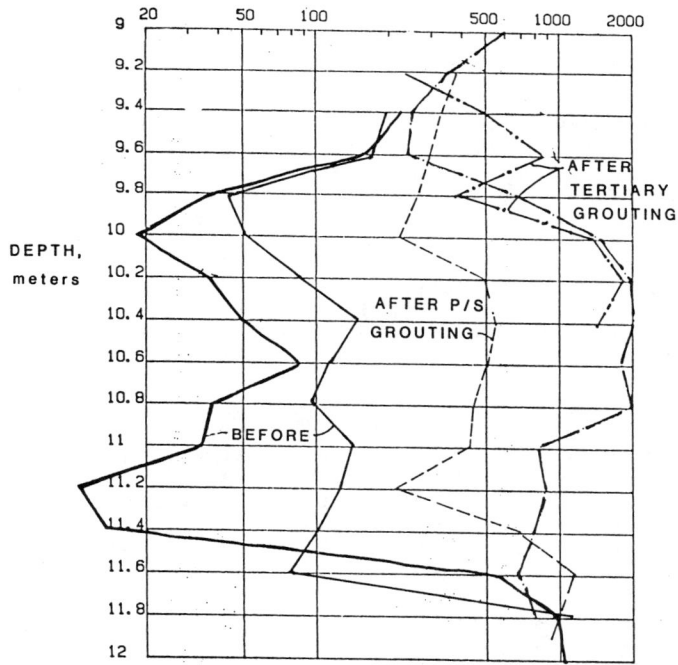

FIGURE 13.

Representative DMT Constrained Modulus Values for Pinopolis Dam Test Plot No. 3 Target Zone.

STEEL CREEK DAM TEST RESULTS

In Steel Creek Dam Test Plot No. 3, the goal was to inject some 7.5 percent total grout volume in the target treatment zones, but the actual grout injection volumes averaged only 5.1 percent and 3.8 percent in the 50-30 feet zone and the 30-20 feet zones, respectively. Net final grout hole spacing was about 12 feet (3.7 m) on centers on a rectangular grid system. A summary of grout quantities at each injection hole for the 40 to 45-foot stage is given in Figure 7.

The following table summarizes the average grout takes in the interior grids per 5-foot treatment depth:

Treatment Depth	Grout Displacement, Percent	
	5' Stage	10' Stage
45 to 50 ft.	5.0	4.8
40 to 45 ft.	4.7	
35 to 40 ft.	5.2	5.4
30 to 35 ft.	5.5	
25 to 30 ft.	4.4	3.8
20 to 25 ft.		

The average heave over the Steel Creek Test Plot No. 3 was 0.12 feet. This translates into a 1.2 percent displacement over the upper 10-foot treatment zone with which most of the heave was associated. If the net heave is subtracted from the average grout take in the upper 10-foot zone of treatment, the resultant volume densification for that zone is 2.6 percent, considerably below the 5.4 and 4.8 percent average displacements obtained in the next two lower 10-foot thick treatment zones. This is reflected in the generally low improvement in the SPT and CPT test data in this upper zone, especially where friction ratios are above 2 percent.

Penetrometer tip resistance and friction ratio values are shown for the 8 to 14 meter zone in Figure 14. Tip resistance values average about a 40 percent increase between the 30-foot and 50-foot depths and show minimal changes above 30 feet. Zones with CPT friction ratios greater than 2 percent generally show much less improvement than zones with friction ratios lower than 2 percent.

Deep lateral displacements were shown in inclinometers 12 to 18 feet outside the treatment area. The inclinometer 12 feet away showed greater than 1.5 inches (3.8 m) lateral displacement away from the injection zone between the 25 to 45-foot depth. This suggests a large radius of displacement away from the injection point, which is more typical of cohesive soils.

The reduced effectiveness of compaction grouting densification at the Steel Creek Dam test site appears to have been related to the use of a grout mix which sand blocked at too low a pressure, the effects of high plastic fines in the treatment zone soils which restricted rapid densification, and the sloping site geometry.

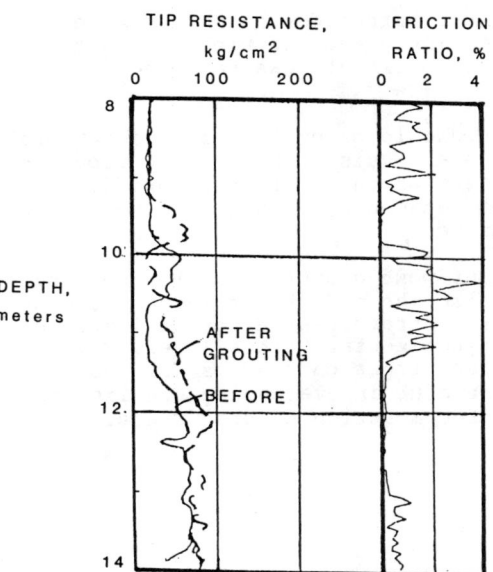

FIGURE 14. Representative Cone Penetrometer Tip Resistance Values for Steel Creek Dam, 40 to 45-Foot Stage.

CONCLUSIONS

Compaction Grouting is a feasible alternative for in-place densification of loose, potentially liquefiable embankment dam foundation support materials. The following guidelines should be considered in planning and performing embankment foundation densification work:

(1) Grout holes should be installed and grouted in increasingly narrower grids, spaced close enough to densify intermediate soils between injection points.

(2) Grout mix designs must provide for high injection pressures without sand blocking, yet must be of low enough plasticity not to permit uncontrolled lensing.

(3) Grout injection criteria should include: (A) grout volume limits for each hole stage that reflect a rational distribution of predicted grout take quantities; (B) site specific injection pressure limits that will result in adequately densified ground but are below the grout mix sand blocking pressure; (C) heave and displacement limits consistent with site geometry and adjacent structures; and (D) cessation of injection when sudden pressure reductions occur.

(4) Embankment overburden depths should be great enough to permit injection pressures high enough to consistently densify loose soils at reasonable hole spacings without heave.

(5) Grout takes should be evaluated by a rational grid distribution analysis with tertiary grout holes located at the centers of the evaluation grids. Grout take percentages between adjacent grids should be compared for consistency.

(6) Evaluation of post-grouting in-situ testing results must recognize the effects of large increases in lateral soils stresses after grout injection. The dilatometer Horizontal Stress Index, K_D, appears to reflect these stress changes as well as density changes and may be useful in evaluating the liquefaction potential of compaction grouted sands.

REFERENCES

Baker, W. H., Cording, E. J., and MacPherson, H.H., "Compaction Grouting to Control Ground Movements during Tunneling", Underground Space, Vol. 7, pp. 205-212, 1983, (Pergamon Press Ltd).

Donovan, N. G., "Site Improvement in a Sensitive Environment", pp. 185-190, Proceedings of the International Conference on In-Situ Soil and Rock Reinforcement, Paris. France, October 1984.

Graf, E. D., Personal communication concerning Kaiser Hospital Site Improvement Grouting, June 1984.

Robertson, P. K., and Campanella. R.G., "The Flat Plate Dilatometer Test for Liquefaction Potential." Soil Mechanics Series No. 79, Department of Civil Engineering. the University of British Columbia, Vancouver, Canada. May 1984.

Schmertmann, J. H., and Baker, W. H ., Subsurface Densification Test Program for St. Johns River Power Park Units 1 & 2, for the Jacksonville Electric Authority, Jacksonville, Florida, July 1983 (Hayward Baker Company).

Warner, J., "Compaction Grouting - The First Thirty Years", ASCE Specialty Conference on Grouting in Geotechnical Engineering, New Orleans, LA. February 1982.

COMPUTER APPLICATIONS IN GROUTING

Michael Demming[*], James L. Rogers[**], Alex Tula[***]

PREVIOUS RECLAMATION GROUT MONITORING PROCEDURES

Prior to 1982, the Bureau of Reclamation had no means or plans available to implement either a computer or electronic monitoring program in dam foundation grouting, even though the technology was available. The operating procedure during grouting was for inspectors on the site to count bags of cement and keep track of the water placed in the grout tubs. This information provided the inspector with the water/cement ratio and provided Reclamation with the bag rate per hour going down the grout hole. The inspector on the foundation kept track of the pressure as indicated on a pressure gauge attached to the header. The water/cement ratio, flow rate, and pressure were recorded on daily forms which were then tabulated on daily, weekly, and monthly reports and drawings.

The usefulness of this data, although valuable, was quite limited. Because of the large number of grout holes drilled on a foundation (approx. 2000), it was almost impossible for one person to review all the daily and weekly reports. The best that could be done was a quick review of the monthly reports. Only if a problem occurred during grouting or after the structure was completed would the grout data be thoroughly analyzed. In most cases, the design for the foundation grouting of a new structure was based on past Reclamation practice and what was remembered about problems encountered during the last grout program.

FIRST GENERATION COMPUTER MONITORING

The first use of computers during foundation grouting was tried by Reclamation on Ridgeway Dam in Colorado. The system was developed by the Hayward Baker Company who was the grouting subcontractor, and was used to monitor flow and pressure. It was very useful in monitoring real time down hole conditions and for predicting fracturing or foundation uplift. The system did, on a secondary basis, produce some reports and half-sized plan and profile drawings. The lack of system dependability, limited capability, nonstandard grout reports, and inadequate drawings for Bureau needs led the Upper Colorado Region and

[*] Chief, Geology Branch, U.S. Bureau of Reclamation, Duchesne, UT.
[**] Engineering Geologist, U.S. Bureau of Reclamation, Duchesne, UT.
[***] Project Manager, GEOMIN Computer Services Corporation, Lakewood, CO.

Note: Authors listed in alphabetical order.

Engineering and Research (E&R) Center to the conclusion that a more reliable, tailored system was needed. Because of problems mentioned, use of the data was limited; however, the Baker Company should have been able to use much of the reporting data for accounting and contract administration purposes.

SECOND GENERATION COMPUTER MONITORING

Staff from the E&R Center Design, Geology, and Construction Divisions, Upper Colorado Regional Office, the Uinta Basin Construction Office, and Montrose Projects Office met on various occasions to determine needs and develop criteria for a complete monitoring system which would be developed for Bureau-wide use. In addition to the requirements to record water/cement ratios, flows, and pressure, it was decided to take full advantage of existing computer technology and have the system compile and print all required Reclamation reports and drawings and provide real-time display of incoming grout data from the headers. Real-time display would report, in color, up to six headers (holes) being grouted and show on a screen graph: (1) bag rate per hour, (2) pressure, (3) flow, (4) flow/pressure, (5) total bag rate for the hole being displayed and during a water pressure test, (6) water take in gallons per minute. The water/cement ratio could be keyed in from the keyboard of the computer or, using nuclear density meters on the pressure side of the grout plant, report directly to the computer the density of the slurry passing through the line, which would then be converted to a water/cement ratio by the software. This incoming data would then be stored on magnetic media for future reports, drawings, and statistical interpretation by both Geology and Design personnel.

As often happens in most engineering organizations, the generation of reports and drawings ultimately requires more manpower and time than the actual construction of the structure. Using the computer in conjunction with state-of-the-art printers and plotters appeared to be an obvious solution in the reduction of time and manpower required to produce the necessary drawings and reports. During meetings with Reclamation personnel, a list of all required field data sheets, report forms, and drawings was compiled and made part of the specifications. In addition, several new reports were discussed and made part of Reclamation requirements. Both size and format, with only minor changes, had to be maintained.

In order for the system to generate required reports and drawings, as well as provide Geology and Design with historical grout data for statistical analysis and comparison, it was necessary that all incoming data be stored on a medium which was not only reliable, but would provide easy access to both field and E&R Center personnel.

In addition to permanent storage, all data collected would have to be compatible with existing Field and E&R Center computer software and hardware and be easily transportable. The Bureau of Reclamation currently has numerous Hewlett-Packard desktop computers in use at the E&R Center, Regional, and Field Offices. A prime requirement was that any system used in grouting must be fully compatible with the existing Bureau-owned software and hardware.

Past experience in procuring new software and hardware has demonstrated that, in most cases, what the user wants the system to do and what happens after the system is installed results in many modifications. The modifications may be a result of expecting too much from the system, poor specifications, a need identified after the system becomes operational, or unexpected "bugs." Whatever the reason for the modification, the software developer must be willing to modify the programs to satisfy the user. This can be accomplished either on a maintenance agreement or through contract modification.

The most often overlooked, but most important, item missing in computer specifications is user and system documentation. If the user cannot easily get in and out of the system (this includes the most casual of operators), frustration takes over and the system will never be used to full capacity. In many cases, the system is abandoned, and the old way is used again because "it works." The importance of easily read and understood manuals cannot be over-emphasized.

If the software has been developed exclusively for, and is owned by, the user, system documentation, including source code and flow charts, must be acquired. The user has no guarantees the software company will be in business after the acquisition is made; or, in many cases, if the programmer has left the company, that the company will support the software. If the software is structured and well documented, the user may be able to contract to another company or programmer for software support.

During our evaluation of the first generation system, it became obvious that regardless of the hardware purchased, it must be able to function in a construction environment. This includes, within reason, dust, moisture, and fluctuating power. Since no equipment is 100% reliable, dependable and rapid maintenance must be available.

PERFORMANCE AND OPERATION

The system developed by Reclamation and GEOMIN was in operational use for approximately two months at the Stillwater damsite; initial operation was delayed due to a late start in the grouting operations, and then grouting was terminated by the contractor due to weather. Most of this time was spent testing and debugging the software, since the system could not be tested in an operational environment prior to actual grout operation at the Stillwater damsite. In addition to debugging the system, two major modifications to the software were made and several pieces of equipment (uninterruptable power supplies) were ordered. These changes were made to include items not considered during initial software development. They include manual entry of data, drawing orientation, and accounting for redrill and backfill grout. In several instances, because of generator power shutdowns, data was lost, and a decision was made to provide for at least a 10-minute power backup to provide for an orderly shutdown of the equipment and prevent corruption of databases. The entire debugging and modification process should be complete prior to spring startup and will be fully tested about one to two weeks after grouting begins. After final testing, the system should be fully operational.

HARDWARE CONFIGURATION

Chief among the concerns when designing the hardware to be used with the system was reliability and maintenance. Past experience had shown that ordinary personal computers do not survive well in a construction environment. It was also felt that on-site maintenance by the manufacturer would be required since prolonged shutdowns were not acceptable. It was recognized that there would be advantages in terms of responsiveness to maintenance needs if as many of the components as possible could be supplied by the same manufacturer. There is a tendency in the computer industry, when components of several manufacturers are combined, for each manufacturer to blame the other companies' equipment when a malfunction occurs. Such conflicts needed to be kept to a minimum.

The basic hardware configuration shown on Figure 1 was accepted as a preliminary design. The two desktop computers (labeled "workstations" in the figure) would operate independently in data acquisition and data processing modes, respectively. Data processing functions would include project setup, grout stage initiation, and report and plot generation. Since the two computers would be functionally identical, some hardware redundancy was provided in case of a machine failure. Leaving one machine dedicated to real-time functions eliminated concerns about interrupt timing and priorities which could potentially result in data loss.

The disk controller is a necessary "black box" when two or more stand alone computers are networked to a common database. It functions much like a "traffic cop" in disk access operations. For mass storage, a Winchester technology disk was specified. Besides providing considerably larger storage volume (up to hundreds of megabytes) than familiar floppy disks, the Winchester-type drives are much more immune from environmental contamination since they are sealed units. For instance, the space between the read/write head and the magnetic media in a floppy disk drive is less than the size of a typical airborne particle of cement dust. The Data Acquisition Unit included the analog-to-digital (A/D) converters, digital voltmeter, and multiplexers for up to 40 channels. Although only 24 channels would be used initially, expansion capability was desired. A graphics printer was to be attached to each workstation to allow dumps of screen graphic displays. For the report generation machine, the printer needed also to be of fairly high speed and high print quality, and have a 220-character print width. A drafting plotter was also required which would be capable of ink-on-mylar plots of standard Bureau of Reclamation drawings (21 in x 36 in (53 cm x 91 cm)).

In selecting the hardware, several criteria in addition to maintainability and reliability were considered. First, it was recognized that the system would be operated by construction inspectors who might be essentially "computer illiterates." Therefore, the computer operating system needed to be either very simple or preferably transparent to the operators. In order to capture and process data on up to 40 channels at 10-second intervals, a fairly high speed central processing unit (CPU) would be needed. Screen graphics needed to be fast (to allow switching between holes in reasonable time) and high

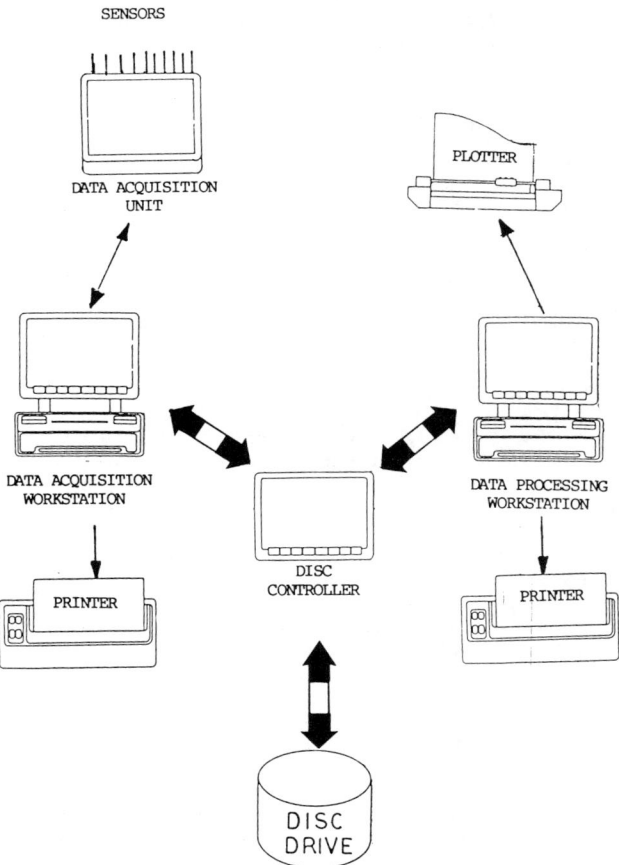

Figure 1. Hardware Configuration

resolution. A color display was preferable for the data acquisition screen display.

After considering several alternatives, the hardware system selected by the Bureau was that recommended by GEOMIN Computer Services which form was simultaneously selected to supply the system software. All hardware components would be supplied by the same manufacturer (the Hewlett-Packard Company) and had been designed for operation specifically in an instrumentation environment. One of the key features for operating in a real-time environment is the facility for time-based and software-based interrupts with programmable priority levels. Both workstations would be Model 9836 desktop computers. These machines use the 16/32 bit Motorola 68000 microprocessor running at 8 MHz as their CPU, but also incorporate four other microprocessors dedicated to input/output (I/O), graphics, and other functions. These considerably speed the throughput of the system. On-site maintenance was available directly from the manufacturer with 24-hour response time. Compatibility with all peripherals was guaranteed by the manufacturer.

SOFTWARE DEVELOPMENT

One of the first questions faced in the development of any software system is what programming language to use. For the selected hardware system, three choices were available: Fortran 77, Pascal, and Hewlett-Packard "Series 200" BASIC (H-P BASIC). H-P BASIC overcomes many of the problems associated with most other versions of BASIC, while retaining programming ease, interpretive debugging and editing capabilities, and user-friendliness of the language. It has extremely fast execution speeds, full numerical operation, local variables and structural constructs normally available only in Fortran; and extensive alphanumeric manipulative facilities. It also possesses simple yet versatile I/O syntax, including communications with the data acquisition peripherals. Finally, the language includes powerful, engineering-oriented graphics commands, which are identical for both the screen display and the drafting plotter.

One of the overwhelming concerns was the time available to develop the system. Only three to four months were allowed in the original schedule between contract award and the required installation date. This for a system which would ultimately include 20 separate program modules and over 20,000 lines of code. Coincidentally, procurement difficulties resulted in the contract not being awarded until the system was functionally completed (the installation date, of course, could not be delayed). Fortunately, GEOMIN had an extensive library of already-developed engineering and geological systems, including instrumentation systems, which had been written using the same series computers and the previously described BASIC language. Many of these systems had already been acquired by the Bureau. Because of timing and budgetary constraints imposed by the Bureau, the system could only be developed by extensive modification of existing software and holding new system development to a minimum. In addition, the Bureau's technical specifications were only three pages in length, making it highly probable that numerous changes would later be required. For these reasons, the system would be developed in "Series 200" BASIC. It should be noted that since the system was developed largely through modification of proprietary systems, and since the Bureau paid a fixed catalog-product price, that copyright and proprietary rights were retained by GEOMIN.

The over-all system design is shown on Figure 2. The "READ/WRITE" indicators show the major function of that element. For instance, the data acquisition facility must periodically read the calibration file, but its principal function is to write the instrument readings to appropriate files. For ease of use, the entire system would be menu driven throughout. To perform a specific activity, the user presses an appropriate special function key. Menus are "tiered" with only the appropriate menu displayed at any given time. Data entry is done primarily through screen forms, similar to familiar spreadsheet programs. The "set up" programs include facilities for generating new projects, calibrating the instruments, inputting new drill hole information, and bringing up new grout stages.

The data acquisition module includes all the real-time activities. These include reading the input channels on 10-second intervals, converting the captured voltages to equivalent engineering units, updating the screen display, and updating the database. It is unacceptable in a real-time environment to have the computer waiting for a keyboard entry from the operator, since this can result in data loss. Therefore, all interaction is done through special function keys with interrupt priorities lower than the data acquisition functions, and through communication files which are written by the data processing workstation and read at software-specified intervals. Presently, 24 channels are read, representing up to six separate grout stages. Each grout stage requires four instrument readings: outbound grout flow; return grout flow; collar pressure; and grout density (from a nuclear density gauge). The screen display includes an alphanumeric display of the data (in engineering units) for each of the six grout stages, and a graphical display of one of the six, selectable by the operator. Note that flow is expressed both in gallons per minute (gpm) of slurry and bags per hour of cement, which is a function of total volume and water/cement ratio. The graphical display shows curves for the past six hours of pressure, flow (gpm), bags per hour, and flow divided by pressure. After each series of 10-second readings is converted to engineering units, it is posted to a rotary file on disk which contains 24 hours of such readings. This overcomes the problem encountered in the original system where a brief power outage can result in the loss of hours of grouting data. Every 15 minutes another rotary file is updated with the average and accumulated totals for that 15-minute interval. This file is maintained for about two months, whereupon the oldest information is overwritten with newer data. Also every 15 minutes, the files for each hole currently grouting is updated by accumulating grout totals and other data for that stage.

When the machine is not otherwise occupied with these tasks, the operator can request display of a different grout stage, modify the display format, or output the screen display to the printer.

Five different reports are generated by the system, all in more or less traditional Bureau of Reclamation formats. These include the

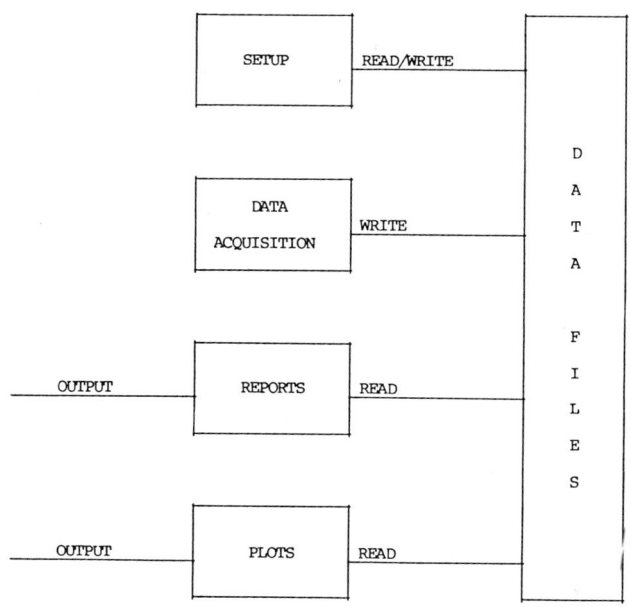

Figure 2. Software System

Drilling Inspector's Report, Grouting Inspector's Report, and three somewhat different summaries of drilling and grouting operations. The system can also create four different drawings: plan and section views of grouting status and grout takes by stage, and contour plots (in plan and section) of grout takes in bags per foot. For both reports and plots, extensive user control is provided of locations, type, date, and other criteria for the data included.

CONCLUSIONS

The system represents a dramatic improvement in the quality of data recovered by the Bureau during grouting operations. This should result in both better managed grouting programs, as well as a better understanding of the grouting process itself. Although the system is billed as a monitoring system, less than 20 percent of the software is dedicated to data acquisition and real-time processing. The balance of the software generates numerous routing drawings and reports which had previously been done laboriously by hand. Of considerable significance, during the five months the system was in service in 1984, not one day was lost to hardware malfunction. The Bureau plans to continue use of the system at Upper Stillwater Dam when grouting operations resume in the spring of 1985, and expand its use to other damsites which will begin construction in the near future.

COMPUTER-ASSISTED GROUTING
EVALUATION SYSTEMS

Leland F. Grant, Member ASCE*

ABSTRACT

Electronic monitoring of cement pressure grouting at Ridgway Dam provides a powerful tool for geologic assessment of drilling and grouting results. Graphic surveillance and analysis of the ongoing grouting operation allow immediate identification of the filling, tightening, and checking phases. With these aids, the trend toward a point of diminishing returns and ultimate refusal of grout take can be skillfully confirmed.

The compiled results of sequentially scheduled primary, secondary, and tertiary holes show trends toward a point of diminishing returns. Grout takes for the primaries respond predominantly in the filling phase, with minor quantities of hole footage in the tightening or checking phase. Grout takes for the secondaries respond with reduced filling phase results and increased amounts of terminal tightening and checking phase characteristics. When the treatment has been monitored and controlled through primary and secondary grouting, the tertiaries show significant terminal responses.

Any substantial divergence from this trend indicates a response to significant bedrock weathering or dissolution features. These anomalous responses can be coordinated with bedrock characteristics projected from the drill hole logs to evaluate the need for quaternary check hole treatment. This ensures the most economical and effective grout treatment possible.

The electronic monitoring data file is also a powerful research tool in the assessment and evaluation of seepage and settlement surveillance during operation of the completed structure. Quick access to the computer file assists in the identification of seepage sources and settlement or uplift pressure events.

INTRODUCTION

The U. S. Bureau of Reclamation (USBR) has developed a computerized electronic monitoring system for collecting, processing, and storing data compiled during cement pressure grouting work.(1) The system has been used during construction of the second stage foundation at Ridgway Dam on the Uncompahge River. A second application of the system is scheduled for the construction of the Brantley Dam foundation on the Pecos River.

*Chief Geologist, Hensley-Schmidt, Inc., Engineers and Managers, 817 Broad Street, Chattanooga, Tennessee 37421.

The system is capable of collecting and processing data to provide significant parameters for controlling the grouting work and assessing the results. One of the most important examples of this capability is the periodic summaries of the grout mix take. The output of these two variables can be processed into graphic displays that can be assessed for ongoing grouting results. Also, the two variables can be processed by mathematical transformations to provide observations and arrays of variables that aid in the positive identification of the point of diminishing returns and grout refusal.

The computer software used in the monitoring programs was specified to provide the equivalent data that could be obtained by using USBR Form 1374: Grouting Inspector's Report. Similar data was recorded manually on the Form 1374 reports for the 1983 test grouting at Brantley Dam. The Brantley test grouting data was used to manually calculate the variables and plot the graphics used to demonstrate the proposed computer processes.

The limitations on space makes it impossible to include in this summary large numbers of drawings and tables to illustrate every aspect of operational control and assessment of grouting results. There are indications of a need to carefully examine the complex features of cement pressure grouting that are involved in grout take refusal. Computer analysis may be capable of filling these and could provide substantial help in solving the problems. Therefore, the essential elements have been addressed with anticipation that other contributions will expand the system.

GROUT TAKE ASSESSMENT

The grouting inspector's reports for the test grouting at Brantley Dam show diminishing trends in grout takes. But without processing these observations, there are no indications of patterns that identify the approach of positive refusal. However, with the aid of computer processing, the grout flow parameters can be identified and refusal characteristics determined. This is accomplished by breaking down the grout takes according to grouting stages and mix changes within each stage.

Graphic Displays

Figure 1 shows the results of this breakdown of grout takes for Primary Hole 71+10 in Section 1 of the Brantley Dam test grouting work. This example is an exhibit that can be manually compiled from the performance data recorded on the form (Form 1374) covering that work. The equivalent computer output data from Ridgway Dam have been examined and they show that similar graphics can be created in the computer and displayed on a video screen during the progress of grouting work.

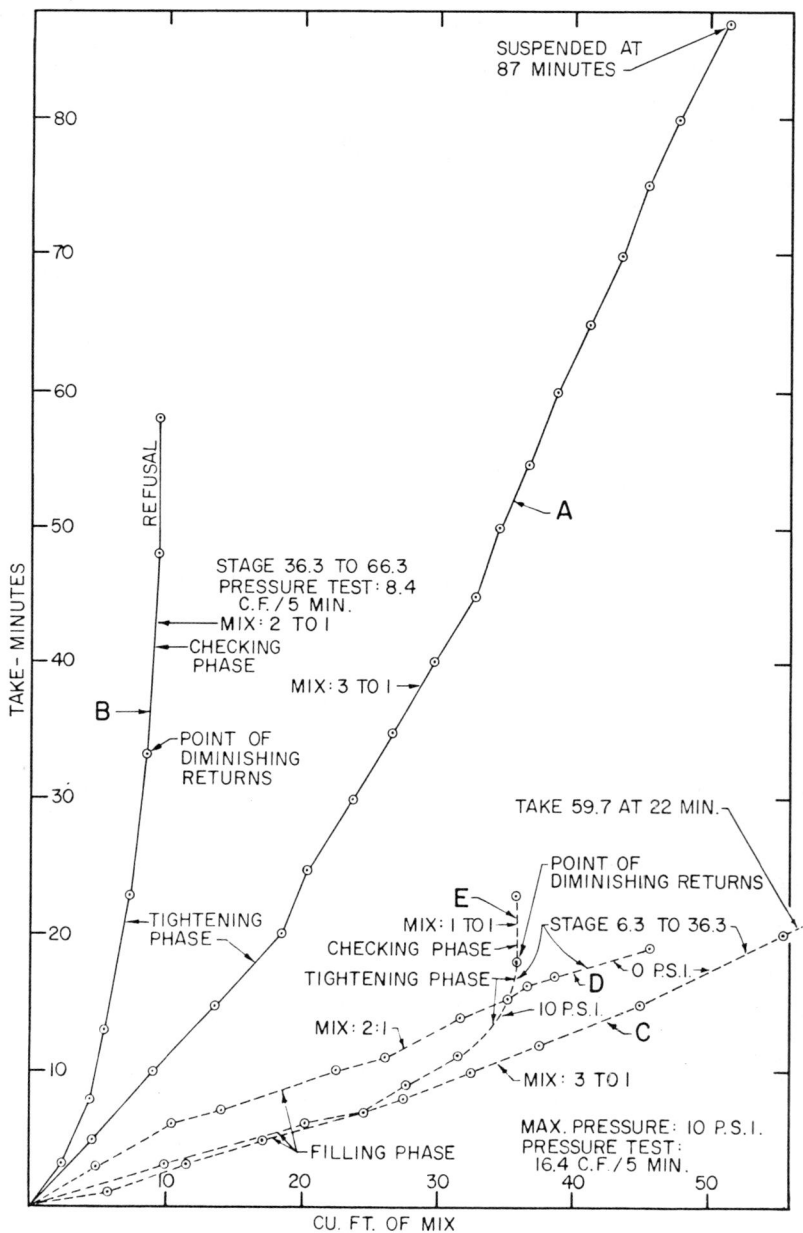

FIGURE 1

The graph on Figure 1 is useful in determining categories of gross flow characteristics of each hole grouted. The curves are plotted as accumulative time versus accumulative cubic feet of grout mix, beginning at zero for each mix used in each stage. This breakdown provides a graphic track of the results of grouting each segment of the work performed while injecting Hole No. 71+10. This separation of the curves into shorter segments permits use of a larger scale which results in quicker and more accurate assessment of the curve deflections. Also, it is possible to plot all segments on a single page or display on a single video screen. With time on the vertical axis, it is possible to scroll the video screen to include longer time periods.

Stage 1

On Figure 1, Curve A is the plot for Stage 1 3:1 grout mix. It shows an irregular but gradual upward deflection caused by diminishing grout flows as internal resistance to the constant pumping pressure (30 psi) increased. This is a typical curve form for a tightening phase response. After 87 minutes of injection, the mix was changed to 2:1 and the resulting Curve B shows greater vertical deflection for an increased tightening phase response. The point of diminishing returns was identified at 33 minutes as the response entered a checking phase. Zero take refusal was recorded between 48 and 58 minutes of pumping time.

Stage 2

Curves C, D, and E are the plots of the accumulative grout mix takes for the three different mixes used in grouting Stage 2 (6.3 to 36.3 feet). The 3:1 mix is represented by Curve C. This injection did not develop pumping pressure, but there are two slight upward deflections in the curve which indicates small increases in resistance to the flows. Curve D is the plot of the 2:1 mix takes which also flowed without pumping pressure. There is a slight downward deflection of the curve at 6 minutes which indicated an increase in flow rate. Curve E represents the 1:1 mix takes which flowed with 10-psi pumping pressure. All of Curves C and D and the first 11 minutes of Curve E indicate a filling phase response in which 140.2 cubic feet of mix was injected. Between 11 and 18 minutes, the acceptance response passed through its tightening phase and entered its checking phase. Refusal at zero take was confirmed at 23 minutes.

COMPUTER VARIABLES

Water Pressure Test Responses

The variables in Table 1 have been compiled from the forms (Form 1374) prepared for Hole No. 71+10 during the Brantley Dam test grouting. In this table, the water pressure test has been entered as a computer observation within the equivalent time frame of 9:35 a.m. to 9:40 a.m. Mathematical transformations of these two time variables provide the variable for the units of time in the interval and the total units of time accumulated during the interval.

TABLE 1

STAGES 36.3 TO 66.3 AND 6.3 TO 36.3

From	To	Minutes	Cumulative	Take	cfm	cfm/LF	Cumulative
A1 Water Test - Stage 36.3 to 66.3 (30 feet)(Date 5-4-83)							
9:35	9:40	5	5	8.4	1.68	0.0560	--
B1 Mix 3/1 - Stage 36.3 to 66.3							
10:00	10:05	5	5	4.4	0.88	0.0293	4.4
10:05	10:10	5	10	4.5	0.90	0.0300	8.9
10:10	10:15	5	15	4.9	0.98	0.0327	13.8
10:15	10:20	5	20	3.4	0.68	0.0227	17.2
10:20	10:25	5	25	3.2	0.64	0.0213	20.4
10:25	10:30	5	30	3.3	0.66	0.0220	23.7
10:30	10:35	5	35	2.7	0.54	0.0180	26.4
10:35	10:40	5	40	2.9	0.58	0.0193	29.3
10:40	10:45	5	45	2.3	0.46	0.0153	31.6
10:45	10:50	5	50	2.5	0.50	0.0167	34.1
10:50	10:55	5	55	2.5	0.50	0.0167	36.6
10:55	11:00	5	60	2.3	0.46	0.0153	38.9
11:00	11:05	5	65	2.3	0.46	0.0153	41.2
11:05	11:10	5	70	2.3	0.46	0.0153	43.5
11:10	11:15	5	75	2.0	0.40	0.0133	45.5
11:15	11:20	5	80	2.0	0.40	0.0133	47.5
11:20	11:27	7	87	3.9	0.56	0.0186	51.4
B2 Mix 2/1 (Date 5-4-83)							
11:27	11:30	3	3	2.5	0.83	0.0278	2.5
11:30	11:35	5	8	1.6	0.32	0.0107	4.1
11:35	11:40	5	13	1.2	0.24	0.0080	5.3
11:40	11:50	10	23	1.8	0.18	0.0060	7.1
11:50	12:00	10	33	1.4	0.14	0.0047	8.5
12:00	12:15	15	48	0.7	0.05	0.0016	9.2
12:15	12:25	10	58	0.0	0.0	0.0000	9.2
A2 Mix (Water Test) - Stage 6.3 to 36.3 (30 feet)(Date 5-4-83)							
12:55	1:00	5	5	16.4	3.28	0.1093	--
B3 Mix 3/1 - Stage 6.3 to 36.3							
1:15	1:16	1	1	5.7	5.7	0.1900	5.7
1:16	1:18	2	3	5.1	2.55	0.0850	10.8
1:18	1:19	1	4	3.2	3.2	0.1067	14.0
1:19	1:20	1	5	3.3	3.3	0.1100	17.3
1:20	1:23	3	8	10.5	3.5	0.1167	27.8
1:23	1:25	2	10	4.6	2.3	0.0767	32.4
1:25	1:27	2	12	4.8	2.4	0.0800	37.2
1:27	1:30	3	15	5.7	1.9	0.0633	42.9
1:30	1:31	1	16	2.2	2.2	0.0733	45.1
1:32	1:36	4	20	10.3	2.58	0.0858	55.4
1:36	1:38	2	22	4.3	2.15	0.0717	59.7

TABLE 1 (Continued)

STAGES 36.3 TO 66.3 AND 6.3 TO 36.3

	Time				Quantities		
From	To	Minutes	Cumu-lative	Take	cfm	cfm/LF	Cumu-lative
B4 Mix 2/1 (Date 5-4-83)							
1:38	1:41	3	3	4.9	1.6	0.0544	4.9
1:41	1:44	3	6	5.7	1.9	0.0633	10.6
1:44	1:45	1	7	3.4	3.4	0.1133	14.0
1:45	1:48	3	10	7.5	2.5	0.0833	21.5
1:48	1:49	1	11	2.5	2.5	0.0833	24.0
1:49	1:52	3	14	7.5	2.5	0.0833	31.5
1:52	1:54	2	16	4.6	2.3	0.0767	36.1
1:54	1:55	1	17	2.6	2.6	0.0867	38.7
1:55	1:59	4	21	8.7	2.2	0.0725	45.4
B5 Mix 1/1							
2:12	2:15	3	3	10.0	3.33	0.1111	10.0
2:15	2:18	3	6	10.6	3.53	0.1178	20.6
2:18	2:19	1	7	3.9	3.90	0.1300	24.5
2:19	2:21	2	9	3.1	1.55	0.0517	27.6
2:21	2:23	2	11	3.6	1.80	0.0600	31.2
2:23	2:26	3	14	3.4	1.13	0.0378	35.1
2:26	2:27	1	15	0.3	0.30	0.0100	35.4
2:27	2:30	3	18	0.5	0.17	0.0056	35.9
2:30	2:35	5	23	0.0	0.00	0.0000	35.9

The total water flow is transformed into a flow rate per minute (cfm) and an acceptance parameter of cubic feet per minute per linear foot (cfm/LF) of hole in the stage. These two parameters are important in the assessment of the grout refusal results of this stage.

Grout Take Response

Based on the water take in the pressure test result, grouting was started in Stage 1 with a selected 3:1 mix (B1) and at a pumping pressure of 30 psi. At the start of injection, the acceptance parameter in cfm/LF was almost one-half the result of the water pressure acceptance flow. But there was a gradual increase for the first 15 minutes when the flow stabilized at 0.0327 and the rate then dropped to 0.0227 in 5 minutes. The variables in the remainder of this column show a slightly diminishing rate of flows but there are several slight increases which indicate typical tightening phase responses.

After 87 minutes of pumping time with 3:1 mix, the mix was changed to 2:1 and the observations for this grouting were recorded in Table 1-B2. The first observation at 3 minutes shows an acceptance rate of 0.0278 (cfm/LF). The acceptance rate uniformly decreased for each of the observation intervals and at 33 minutes the take dropped below the point of diminishing returns and entered its checking phase response. Between 48 and 58 minutes, the grout take was 0.0 cfm and the acceptance rate was 0.0000 cfm/LF which indicates a valid refusal at grout mix saturation of the open weathered fractures.

Stage 2 water pressure testing and grout mix injection results are recorded in Tables 1-A2, -B3, -B4, and -B5. Table 1-A2 shows a total take of 16.4 cubic feet of water, a pumping rate of 3.28 cfm, and an acceptance rate of 0.1093 cfm/LF. Table 1-B3 shows that 59.7 cubic feet of 3:1 mix was injected in the first 22 minutes. The allowable pumping pressure was 10 psi, but no significant pressure was noted on the header gauge. The acceptance rate varied between 0.1900 and 0.0717 cfm/LF with no positive diminishing trend. This indicates a filling phase response.

Table 1-B4 shows the observations for 21 minutes of 2:1 mix injection at no significant pressure on the header gauge. The irregular acceptance flows indicated the grouting was in a filling phase and there was only a slight increase in benefits for the 2:1 mix injection.

Table 1-B5 shows the observations for 23 minutes of 1:1 mix injection. The stiffer mix required pumping at the maximum allowable pressure of 10 psi. At the start of grouting with this mix, the acceptance flow was 0.1111 cfm/LF and it increased to 0.1300 cfm/LF at 7 minutes. At 9 minutes the acceptance flow declined to 0.0517 and leveled off at 0.0600 at 11 minutes as the injection changed to a tightening phase. The tightening phase ended at 18 minutes as the checking phase started with the zero take refusal. This response indicated rapid expulsion of trapped groundwater and grout saturation of the open weathered fractures.

SECONDARY HOLE RESPONSES

The comprehensive grouting described above was compared with the results of secondary hole treatment at Station 70+90. This example is at the mid-point between the above 71P10 and one other primary hole at Station 70+70. Both of these holes were grouted before the drilling of Secondary Hole 70+90 was started. Table 2-A1 shows that the water pressure test for Stage 1 (37.0 to 67.0) was too tight to justify a separate stage of grouting which indicates substantially less weathering in this part of the bedrock mass. Due to this low test result, the two stages were combined and grouted as one 60-foot stage. The water pressure test result for these combined stages was 1.4 cubic feet of total water take, 0.28 cfm flow, and 0.0093 cfm/LF acceptance rate. (See Table 2-A2.)

The graphic displayed on Figure 2-A shows the point of diminishing returns at 15 minutes of pumping time. The 1.8 cubic feet of mix that was injected in the first 15 minutes completed the tightening phase. The acceptance rate was 0.0053 cfm/LF for the first 5 minutes, dropped to 0.0013 for the next 10 minutes, and

COMPUTER-ASSISTED GROUTING   139

leveled off at 0.0010 cfm/LF for the next 20 minutes. This expulsion of trapped water reduced the acceptance rate to the seepage rate of water and prevented a zero take refusal at the point of diminishing returns. In this case there was approximately 40 minutes of extended pumping time for the checking phase.

Table 2-B1 shows the observations for 60 minutes of pumping time in which the total take was 3.9 cubic feet of 5:1 mix. The volume of the drill hole in the 60-foot stage is approximately 4 cubic feet. This means that most of the grouting effort was spent in expelling the trapped water in the hole.

TABLE 2

STAGE 7.0 TO 67.0

Time				Quantities			
From	To	Minutes	Cumu-lative	Take	cfm	cfm/LF	Cumu-lative
2-A1 Mix (Water Test) 37.0 to 67.0 (30 feet)							
10:38	10:43	5	5	0.2	0.04	0.0013	--
2-A2 Mix (Water Test) 7.0 to 67.0 (60 feet)							
11:00	11:05	5	5	1.4	0.28	0.0047	--
2-B1 Mix 5/1							
11:35	11:40	5	5	1.0	0.20	0.0033	1.0
11:40	11:50	10	15	0.8	0.08	0.0013	1.8
11:50	12:00	10	25	0.6	0.06	0.0010	2.4
12:00	12:05	5	30	0.3	0.06	0.0010	2.7
12:05	12:10	5	35	0.3	0.06	0.0010	3.0
12:10	12:20	10	45	0.4	0.04	0.0007	3.4
12:20	12:30	10	55	0.3	0.03	0.0005	3.7
12:30	12:35	5	60	0.2	0.04	0.0007	3.9

TERTIARY HOLE RESPONSE

Water Pressure Tests

This example of tertiary hole grouting assessment is related to the primary and secondary holes considered above. This hole was drilled at the mid-point and 10 feet distant from each of the above holes. Its drilling and grouting were performed after all work on the primary and secondary holes were completed.

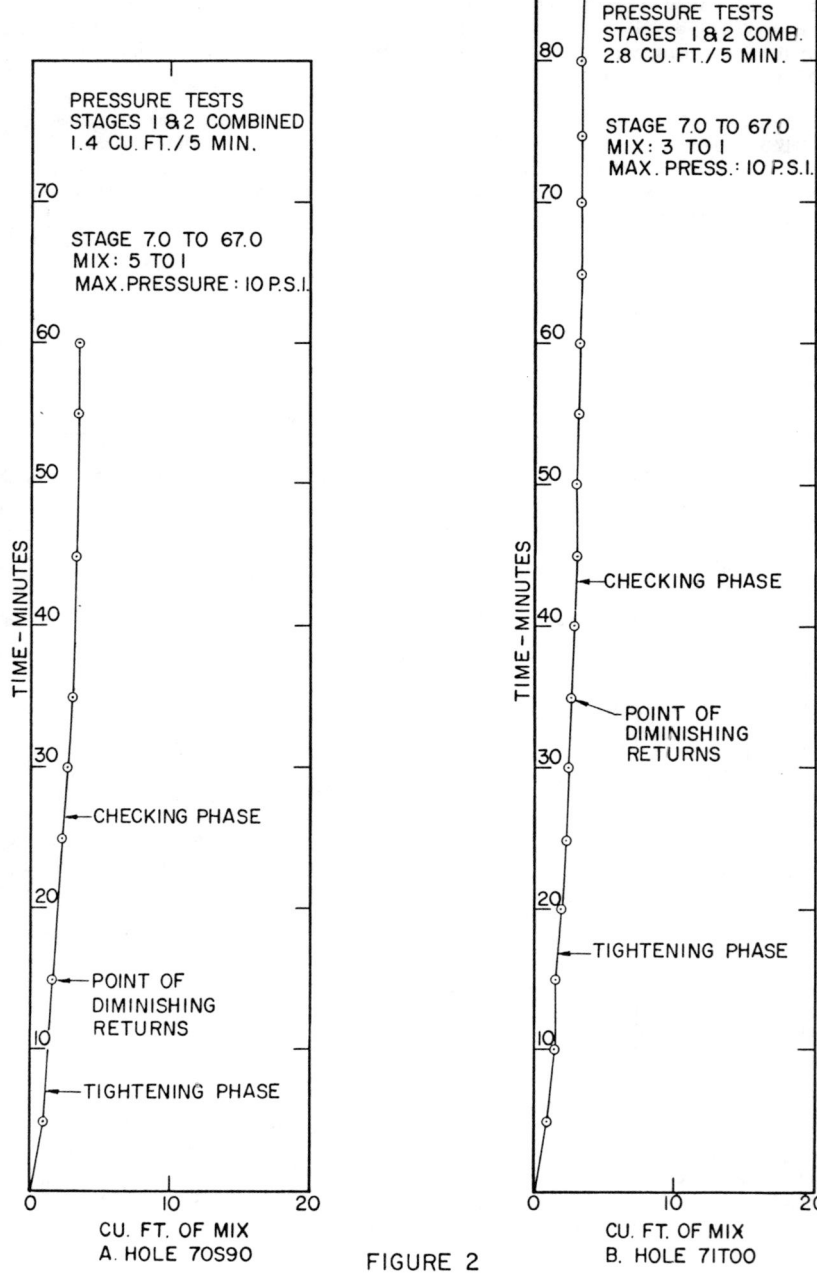

FIGURE 2

Tables 3-A1 and -A2 show the water pressure testing results. Stage 1 at depths of 37.0 to 67.0 shows a total take of 0.1 cubic foot for 5 minutes. This has been transformed into 0.02-cfm pumping flow and 0.0007 acceptance flow for the 30 feet of hole in the stage. With this low take, the stage was deleted and the entire 60 feet of hole was grouted as one stage. Pressure testing for the full 60-foot depth showed 2.8 cubic feet total take, 0.56 cfm, and an acceptance flow of 0.0093 cfm/LF.

Table 3-B1 shows the observations for 85 minutes of pumping time in which the total take was 3.8 cubic feet of 3:1 mix. The volume of the drill hole in the 60-foot stage is approximately 4 cubic feet. This means that most of the grouting effort was spent in expelling the trapped water in the hole. In this case, a terminal refusal of zero take was confirmed at the 85-minute pumping time.

TABLE 3

STAGE 7.0 TO 67.0

Time				Quantities			
From	To	Minutes	Cumulative	Take	cfm	cfm/LF	Cumulative
3-A1 Mix (Water Test) - Stage 37.0 to 67.0 (30 feet)							
1:10	1:15	5	5	0.1	0.02	0.0007	--
3-A2 Mix (Water Test) - Stage 7.0 to 67.0 (60 feet)							
1:30	1:35	5	5	2.8	0.56	0.0093	--
3-B1 Mix 3/1							
2:00	2:05	5	5	0.8	0.16	0.0027	0.8
2:05	2:10	5	10	0.5	0.10	0.0017	1.3
2:10	2:15	5	15	0.3	0.06	0.0010	1.6
2:15	2:20	5	20	0.3	0.06	0.0010	1.9
2:20	2:25	5	25	0.3	0.06	0.0010	2.2
2:25	2:30	5	30	0.3	0.06	0.0010	2.5
2:30	2:35	5	35	0.1	0.02	0.0003	2.6
2:35	2:45	5[1]	40	0.2	0.04	0.0007	2.8
2:45	2:50	5	45	0.1	0.02	0.0003	2.9
2:50	2:55	5	50	0.1	0.02	0.0003	3.0
2:55	3:00	5	55	0.2	0.04	0.0007	3.2
3:00	3:05	5	60	0.1	0.02	0.0003	3.3
3:05	3:10	5	65	0.1	0.02	0.0003	3.4
3:10	3:15	5	70	0.1	0.02	0.0003	3.5
3:15	3:20	5	75	0.1	0.02	0.0003	3.6
3:20	3:25	5	80	0.2	0.04	0.0007	3.8
3:25	3:30	5	85	0.0	0.00	0.0000	3.8

[1] 5 minutes of down time due to compressor.

The graphic displayed on Figure 2-B shows the point of diminishing returns at 35 minutes of pumping time. The 2.5 cubic feet of mix that was injected in the first 35 minutes completed the tightening phase. The acceptance rate was 0.0027 cfm/LF for the first 5 minutes, declined to 0.0017 at 10 minutes, and 0.0010 at 15 minutes where it leveled off and declined again at the point of diminishing returns. After 45 minutes pumping at acceptance rates ranging between 0.0003 and 0.0007, the grout take refused at zero between 80 and 85 minutes. This confirms terminal grouting for this part of the curtain.

SEQUENTIAL GROUTING

Sequentially scheduled grouting of primary, secondary, and tertiary grout holes shows trends toward a point of diminishing returns.(2) Grout takes expressed as cubic feet of mix accepted per linear foot of hole (cf/LF) can be classified and organized to help identify ground characteristics and to track changes in these characteristics as the treatment work progresses.

Grout takes for the primary holes are substantial which indicates a predominance of filling phase responses. Usually there are only minor quantities of hole footage in the tightening or checking phases. Grout takes for the secondary holes respond with reduced filling phase results and increased amounts of tightening or checking phase characteristics. In cases where the primary and secondary treatment have been comprehensively controlled, the tertiary responses are significantly terminal.

Data Observations

Table 4 is the data summary for the nine grout holes treated in the test grouting work for Section 1 at Brantley Dam. These observations and variables have been compiled from the drilling, water pressure testing, and grouting quantities required to perform the test grouting.

The observations and variables in Table 4 have been transformed to simulate the forms required for computer analysis. The hole numbers have been derived from the curtain line station with P, S, or T substituted for the plus symbol. The drilling variable is the total number of feet included in the grout stages. The water take is the total cubic feet injected in the specific 5-minute test periods. This value was divided by 5 for cubic feet per minute and by the total footage of drill hole grouted to obtain the water acceptance parameter of cfm/LF.

The 0.0826-cfm/LF water take and 3.360 cubic feet per linear feet grout take in Primary Hole 71P10 indicate penetration of moderately weathered bedrock. The lower water takes for Primary Holes 70P30 and 70P70 indicate slight to no weathering. The substantially lower water and grout takes for the two secondary holes indicate significant filling of fewer weathered bedrock fractures at these secondary locations. The 0.0100-cfm/LF water take and 0.235 cubic foot per linear foot grout take for Tertiary Hole 70T60

COMPUTER-ASSISTED GROUTING   143

indicate penetration of an isolated weathered fracture. The terminal response identified in its treatment analysis and assessment confirms grouting closure of this area. Also, the terminal responses identified for all secondary and tertiary holes in the test sections confirm effective completion.

TABLE 4

DATA SUMMARY
TEST SITE NO. 1
(STATION 70+30 TO 71+10)

		Water Take		Grout Take	
Hole No.	Drilling (L.F.)	C.F./ 5-Minute	C.F./ Minute/ Foot	CF Mix	C.F./Foot
Primary Injection					
70P30	60.0	5.7	0.0190	38.7	0.645
70P70	60.0	9.0	0.0300	16.5	0.275
71P10	60.0	24.8	0.0826	201.6	3.360
3 Holes	180.0	39.5	0.0439	256.8	1.427
Secondary Injection					
70S50	60.0	0.7	0.0023	2.6	0.043
70S90	60.0	1.4	0.0047	3.9	0.065
2 Holes	120.0	2.1	0.0035	6.5	0.054
Tertiary Injection					
70T40	61.0	0.1	0.0003	2.5	0.041
70T60	62.0	1.5	0.0100	14.6	0.235
70T80	60.0	2.9	0.0097	4.2	0.070
71T00	60.0	2.8	0.0093	3.8	0.063
4 Holes	243.0	7.3	0.0060	25.1	0.103
9 Holes	543.0	48.9	0.0900	288.4	0.531

The graphic display of the water testing parameters on Figure 3 shows substantial saturation of the primary holes with the grout mix and a sharp drop in these acceptance parameters for the secondary holes. The curve for the east test curtain segment, which shows the highest grout acceptance in the primaries, refused at the point of diminishing returns with takes for the secondary and both tertiary holes barely large enough to fill the drill hole with grout mix. In the west segment, primary and secondary takes were both lower, but Tertiary Hole 70T60 accepted 10.6 cubic feet of 3:1 mix up to the point of diminishing returns and an additional 4 cubic feet in 45 minutes. Therefore, both segments are at terminal acceptance and additional grouting is not feasible and may be harmful.

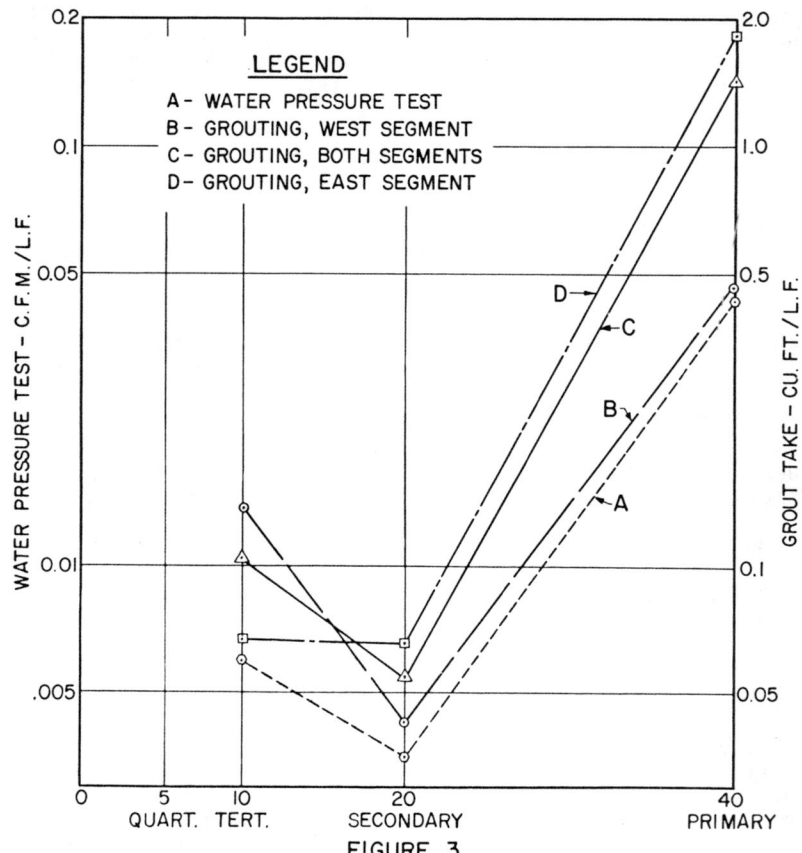

FIGURE 3

Hole 71P10 was water tested and grouted in two separate 30-foot stages with Stage 1 between depths of 36.3 and 66.3 feet took 51.4 cubic feet of 3:1 mix and 9.2 cubic feet of 2:1 mix and refused at zero take for 10 minutes. Stage 2 between depths 6.3 and 36.3 took 59.7 cubic feet of 3:1 mix, 45 cubic feet of 2:1 mix, and 35.9 cubic feet of 1:1 mix and then refused at zero take for 5 minutes. The total take of 201.6 cubic feet of mix for both 30-foot stages was 69.9 percent of the 288.4 cubic feet of mix injected in all nine holes. The other eight holes were grouted as single, combined stages of 60 or more feet, and 86.8 total cubic feet of mix was injected in the 483 linear feet in these holes.

Brantley Damsite is situated on the facies change caused by the El Capitan Reef, a regional aquitard that influences the groundwater movements through this part of the Pecos River Valley. The test grouting program conducted in 1983 was intended to determine and assess the grouting parameters for its treatment during foundation construction. Exploratory core holes located in the area indicate that bedrock in the subsurface consists of interbedded dolomites, siltstone, and thin shale members. All of these rock types have dense lithologies and exceptionally low intergranular porosities in masses of sound and unweathered bedrock. Groundwater seepage and channel flows are developed along bedding planes and joint fractures. Within limits of this part of the grout curtain, bedding plane lineation is essentially horizontal and joint lineation is almost vertical and trends normal to the curtain line.

Table 5 is a computer analysis and assessment of the water tightness characteristics of the Stage 1 grout holes in Test Section 1. In the "Hole Number" column, the testing and grout injection sequence are designated by the letters P-T-S for primary, secondary, or tertiary. The Stage 1 columns identify stage depths and thickness for each hole. The three water columns show the total take for the standard 5-minute test and the acceptance parameters of cfm and cfm/LF of hole.

TABLE 5

WATER TEST PARAMETERS - STAGE 1
TEST GROUTING SECTION NO. 1

Hole Number	Stage 1			Water Takes		
	From	To	Thickness	5-Minute Test	cfm	cfm/LF
70P30	35.5	65.5	30	0.2	0.04	0.0013
70T40	37.0	67.0	30	0.2	0.04	0.0013
70S50	37.0	67.0	30	0.2	0.04	0.0013
70T60	37.0	69.0	32	0.3	0.06	0.0019
70P70	35.0	65.0	30	0.0	0.00	0.0000
70T80	37.6	67.6	30	0.1	0.02	0.0007
70S90	37.0	67.0	30	0.2	0.04	0.0013
71T00	37.0	67.0	30	0.1	0.02	0.0007
71P10	36.3	66.3	30	8.4	1.63	0.0560

The water take parameters for Primary Hole 70P30 indicate that terminal conditions and treatment as a separate stage were not feasible. The water take for Hole 70P70 was zero; it could not accept grout. The water take for Hole 71P10 was 8.4 cubic feet for the 5-minute water test for Stage 1. The curves for this grouting on Figure 1 show tightening and checking phase responses. This indicates occurrence of slight to unweathered bedrock for the western two-thirds of the rock mass and slight to significant weathering for the eastern one-third. Both secondary holes took 0.2 cubic feet for the water pressure tests. During the grouting of Stages 1 and 2 combined, these holes responded as extended checking phases with takes of 5:1 mix too low to completely fill the grout hole. Except for 71T00, the tertiary holes responded with low water takes and extended checking phase grouting in their injection. At Hole 70T80, which is located adjacent to the weathered area penetrated by Hole 71P10, refusal occurred at zero acceptance of 3:1 mix, with the total take of 3.8 cubic feet, which is approximately equal to the volume of the drill hole grouted.

This unusual number of terminal water pressure test and grout take results shows that this part of the grout curtain penetrates the Azotea aquitard, thus tying the curtain to this natural barrier. The higher takes in 71P10 indicate deeper weathering, and the grouting in this test section indicates that weathering effects are confined to the Stage 2 level. The weathering system may extend deeper in areas to the east. This condition needs to be assessed and addressed when treatment of that area is undertaken.

SUMMARY AND CONCLUSIONS

The electronic monitoring of cement grouting tested at Ridgway Dam is capable of generating and recording important grouting data for use in assessment and evaluation of treatment results. Computer processing and analysis of the electronic monitoring system or the Form 1374 data are powerful aids in the control of the effects of foundation treatment.

Computer generated graphics and observations consisting of array variables provide for rapid and accurate determinations of grout refusal and the point of diminishing returns. This is important for fine-tuning ongoing grouting work. Also, positive identification of the occurrences of filling, tightening, and checking phases confirms completion and effectiveness of the grout injection.

Computer processing of the data from sequentially treated holes tracks the changes in water pressure tests and grout takes which respond to the effects of bedrock weathering and dissolution. This tracking can be coordinated with the geologic data obtained from exploratory and grout hole drilling to assess the changes in ground characteristics as the sequence of hole treatment progresses. This provides for skillful selection of needed quarternary closure holes.

The data files accumulated by the monitoring computer are valuable research tools for assessment and evaluation of grouted foundation performance during operation and maintenance of treated structures. Many of the seepage, settlement, or uplift events that are detected in the early stages of project operation are related to anomalous responses in the foundation treatment. These anomalous responses need to be considered in the planning of remedial treatment.

It is feasible to develop and write the necessary subprograms and subroutines needed for assessing and controlling grouting operations. This will provide more effective and economical treatment. Also, the computer systems can be adjusted to conform with existing

TABLE AD-1

STAGES 30.0 TO 60.0 AND 60.0 TO 90.0

Time				Quantities			
From	To	Minutes	Cumu-lative	Take	cfm	cfm/LF	Cumu-lative
A3 Water Test - Stage 3			60.0 to 90.0 (30 feet)				
02:40	02:45	5	5	0.7	0.14	0.0047	--
A4 Water Test - Stage 4			30.0 to 90.0 (60 feet)				
03:02	03:07	5	5	14.7	2.94	0.0490	--
B5 Mix 4/1 - Stage 30.0 to 90.0							
04:04	04:20	16	16	33.6	2.1	0.0350	33.6
04:20	05:20	60	76	84.0	1.4	0.0233	117.6
05:20	05:30	10	86	14.0	1.4	0.0233	131.6
05:30	05:45	15	101	21.0	1.4	0.0233	152.6
05:45	06:00	15	116	21.0	1.4	0.0233	173.6
06:00	06:24	24	140	36.0	1.5	0.0250	209.6
06:24	06:44	25	165	32.5	1.3	0.0217	242.1
06:49	07:29	40	205	44.0	1.1	0.0183	286.1
B6 Mix 3/1 - Stage 30.0 to 90.0							
07:29	07:40	11	11	12.1	1.1	0.0183	12.1
07:40	08:08	28	39	33.6	1.2	0.0200	45.7
08:08	08:27	19	58	22.8	1.2	0.0200	68.5
08:27	08:41	14	72	14.0	1.0	0.0167	82.5
08:41	09:16	35	107	42.0	1.2	0.0200	124.5
09:16	09:33	17	124	20.4	1.2	0.0200	144.9
09:33	09:54	21	145	25.2	1.2	0.0200	170.1
09:54	10:09	15	160	19.5	1.3	0.0217	189.6
10:09	10:40	31	191	31.0	1.0	0.0167	220.6
10:40	11:05	25	216	30.0	1.2	0.0200	250.6
11:05	11:34	29	245	49.3	1.7	0.0283	299.9
11:34	12:02	28	273	36.4	1.3	0.0217	336.3
12:02	12:58	56	329	72.8	1.3	0.0217	409.1
B7 Mix 2/1 - Stage 30.0 to 60.0							
12:58	13:29	31	31	31.0	1.0	0.0167	31.0
13:29	13:54	25	56	27.5	1.1	0.0183	58.5
13:54	14:30	36	92	34.3	0.9	0.0150	92.8
14:30	15:00	30	122	17.5	0.7	0.0117	110.3
15:00	15:30	30	152	20.5	0.7	0.0117	130.8
15:30	16:00	30	182	30.0	1.0	0.0167	160.8
16:00	16:30	30	212	15.0	0.5	0.0083	175.8
16:30	17:00	30	242	6.0	0.2	0.0033	181.8
17:00	17:30	30	272	15.0	0.5	0.0083	196.8
17:30	18:00	30	302	9.0	0.3	0.0050	205.8
18:00	18:30	30	332	6.0	0.2	0.0033	211.8
18:30	19:00	30	362	6.0	0.2	0.0033	217.8
19:00	19:30	30	392	6.0	0.2	0.0033	223.8
19:30	20:00	30	422	3.0	0.1	0.0017	226.8
20:00	20:35	35	457	3.0	0.1	0.0017	229.8

geologic conditions at each foundation site and the grouting specifications required to successfully grout any given foundation. This is justification for the use of computer-assisted systems for assessment and control.

ACKNOWLEDGEMENTS

The data used in developing this study was obtained from sources within the USBR. The drilling and grouting quantities used to compute the illustrated graphics and observations of array variables were from actual test grouting work. The geologic staff at the Pecos River Projects provided detailed and comprehensive data on the test grouting treatment of Brantly Damsite. This was a substantial contribution in compiling the graphics and variables for tracking the grout mix takes and identification of refusal. The report on electronic monitoring at Ridgway Dam provided information on the capability of the computer systems and data output. The writer is notably obligated to engineering and geologic personnel in the Denver Engineering and Research Center, the Southwestern Regional Office, and the Ridgway Dam Project Office for their help and patience in explaining USBR grouting methods.

APPENDIX A

(1) Davidson, Luther, Electronic Monitoring of Cement Pressure Grouting at Ridgway Dam (U. S. Bureau of Reclamation: Denver, Colorado), 1984.

(2) Grant, Leland F., "Concept of Curtain Grouting Evaluation," Journal of the Soil Mechanics and Foundations Division, ASCE, Vol. 90, No. SM1, Proc. Paper 3775, January, 1964, pp. 63-92.

ADDENDUM

The observations and variables compiled in Table AD-1 have been manually generated from the data recorded on the 1374 forms for Primary Hole 25+30.9-15.0 feet u.s. in the Ridgway Dam, left abutment grout curtain. The table tracks grout injection results in two combined stages that include water pressure tested stages between 30.0 to 60.0 and 60.0 to 90.0. The treatment involved 16 hours and 31 minutes of pumping time to place 907 cubic feet of grout mix in the bedrock foundation. A total of 38 summary observations of array variables were recorded for the table.

The time interval between observations ranged from 11 to 60 minutes, with 26 of the time frames covering 20 to 40 minutes. The cumulative time and mix take variables provide adequate data for legible graphics of the grouting responses. The water pressure take variables can be related to those for grout mix takes to confirm valid refusal and assess results of the treatment in sequentially grouted holes. The data on stage depths and diminsions can be related to the total grout mix takes and used to determine effectiveness of the work, and when coordinated with detailed geologic data on bedrock fracturing and weathering, the relationship with natural groundwater barriers can be determined.

ACOUSTIC EMISSION MONITORING OF GROUT MOVEMENT

by

Robert M. Koerner[1], Richard N. Sands[2] and James D. Leaird[3]

Abstract

The detection and monitoring of any subsurface flow phenomenon, such as grouting is a significant and largely unsolved problem area. As a nondestructive testing technique, the acoustic emission (AE) method seems to be a likely candidate technique for application to this problem.

This paper describes the AE method, in general, and in particular the AE system which holds significant promise in this regard. Used was a multichannel AE system which can eventually source locate the grout stimulated emissions in three dimensions and in real time on a CRT screen. A series of laboratory tests, where equipment selection and proper system tuning were made, are described. Two field tests where chemical and cement grouts were injected were also monitored with positive results, indicating the technical feasibility of the method. Computer software has been developed and, is essentially ready for proof testing in the field. Finished graphics work is not finished as of this writing.

Introduction

The injection of grout materials (soils, cements, chemicals, etc.) into soil or rock is a time honored practice which has resulted in many successfully completed projects. Furthermore, there appears to be a sufficient body of knowledge in the practice to know under what situations grouting can be effective and to what degree. The open literature on the subject is indeed voluminous, with case histories being a common vehicle to present information.

There are, however, numerous areas of concern where some additional insight into grout flow patterns and definitive location of the grout front would be of great interest. This is particularly the case for low viscosity grouts in highly porous or fractured media. Here travel paths eminating from the injection point can follow long and tortious random patterns. The obvious investigative method to see what these patterns are is to dig test pits, bore holes, etc., and to physically examine the area. This, of course, is rarely possible to the degree necessary for anywhere near to a complete identification. Needed is a continuous profiling method which does not disturb the in-

---
[1] H. L. Bowman Professor of Civil Engineering, Drexel University, Philadelphia, PA 19104
[2] Graduate Research Assistant, Drexel University
[3] Engineer, AET Corp., Sacramento, CA 95815

situ conditions. Thus nondestructive testing (NDT) methods offer an attractive alternate. While there are many candidate NDT methods available, this paper focuses on the acoustic emission (AE) method.

Acoustic Emission Overview

Acoustic emissions are sounds generated within a material which has been subjected to some type of nonequilibrium disturbance. Sometimes these sounds are audible, but more often they are not, due to their low amplitude or high frequency, or both. A piezoelectric sensor is used as a pickup to detect the acoustic emissions. This sensor produces an electrical signal proportional to the amplitude of sound or vibration being detected. The signal is then amplified, filtered, and counted or recorded in some quantifiable manner. Unwanted machine and environmental noise are electronically filtered from the signal or separately quantified and subtracted from the test results. The counts, or recordings, of the emissions are then correlated with the basic material behavior to empirically determine the relative stability of the situation being evaluated. Usually, if no acoustic emissions are present, the situation is in equilibrium and thus stable. If, however emissions are observed, a nonequilibrium situation exists which may eventually lead to failure. It should be mentioned at the outset that the technique is not new, with rock mine monitoring beginning in the 1930's. Only its application to geotechnical engineering problems is relatively new. Many projects have been described in the Proceedings of three Conferences held at the Pennsylvania State University in 1975, 1978 and 1981, see Hardy and Leighton.[1]

The components of an acoustic emission monitoring system consist of a wave guide (to bring the signals from within the soil or rock mass to a convenient monitoring point); sensor (geophone, accelerometer, hydrophone or transducer to convert the mechanical wave to an electrical signal); preamplifier (to amplify the signal if long cable is being used); filters (to eliminate undesirable portions of the signal); and a quantification system. The quantification system is some form of counter and is usually accomplished on the basis of AE ringdown counts or AE event counts. Ringdown counts are merely the total number of times the AE signals cross an arbitrarily set threshold level. On the other hand, event counts (as used in this study) are individual bursts of AE activity during a preset time interval. The variations of equipment configuration can obviously be great which leads to many possible systems. Currently a number of companies are making single channel AE systems for geotechnical use and some are making multi-channel systems. The option also exists of making a personally designed hybrid system. The general area of AE use in geotechnical work has recently been reviewed by Hardy[2] and Koerner, et al.[3]

AE Monitoring of Flow Phenomena

Previous work with monitoring seepage through earth dams indicated that such disturbances are emittive enough to be sensed by the technique within limitations of background noise, distance from the source, etc. (Koerner, et al.[3]). Separate studies on grout flow monitoring also were positive in that the emissions did indeed occur and it also appears as though hydrofracture pressure could be predicted from changes in the AE rate curve (Koerner, et al.[3]). With these thoughts in mind, work has been directed toward source location of such activities. More specifically, this work is directed at the three dimensional, real-time location of seepage, grouting, and hydrofracturing involved in subsurface flow situations. Both soil and rock masses are candidates for use of the technique.

The key element of the instrumentation is a multi-channel AE pickup system. Required are at least five pickup sensors deployed in the vacinity of the disturbance which stimulate seperate acceptance gates in the detection system. The system distinguishes between arrival times of the disturbances and determines time differences as the wave travels between each pickup sensor. These "Δt" values are then used in a preprogrammed microcomputer for calculation of the actual source in three dimensions referenced to the locations of the pickup sensors. The theory is based on a triangulation algorithm as follows: (Koerner, et al.[4])

$$d_i = [(x-a_i)^2 + (y-b_i)^2 + (z-c_i)^2]^{\frac{1}{2}} \tag{1}$$

where

a_i, b_i, c_i = coordinates of ith sensor
d_i = distance from source to i^{th} sensor
x, y, z = coordinate of source

furthermore, d_i, the distance that the P-wave travels is:

$$d_i = V_i(t_i - t_o) \tag{2}$$

where

V_i = velocity of sound in the direction from the source to the ith sensor
t_i = time at the ith sensor
t_o = time at the source,

therefore,

$$V_i(t_i - t_o) = [(x-a_i)^2 + (y-b_i)^2 + (z-c_i)^2]^{\frac{1}{2}} \tag{3}$$

Equation 3, when reduced to a series of equations and appropriate measured times and known a, b, c values at the sensors, can be solved simultaneously for the locations of the AE source.

Past Work on Grout Monitoring

The ultimate goal of this project is to sense the subsurface noises created by grout flowing in the soil or rock mass and then to calculate and display the source information. The display will be on a CRT screen in three dimensions and in real time. The general configuration is shown in Figure 1. Here it is shown that the source travels to the pickup sensors in the form of radially propagating AE events. The Pioneer 5000 system (which can handle up to eight channels) calculates the "Δt's" and together with the velocity of propagation of sound in the soil and rock involved are input into microcomputer 1 for calculation of the X, Y, Z coordinates of the source. These coordinates are referenced to an arbitrary datum to which the pickup sensors were originally referenced. Details of this program were described in the previous section. Although not currently operational, this coordinate source location data is now fed to a second microcomputer which plots and stores each data point as it is received. As data is continually plotted, a trajectory of activity is developed which (presumably) is the grout front as it is being developed.

Previously reported progress toward this goal has centered around the following activities.

- Large scale laboratory tests used to measure characteristics of the AE event, such as amplitude, rise time, frequency and velocity of propagation in different media.

- Background noise assessment and feasibility studies on a chemical grouting project in Pittsburgh, Pennsylvania.(4)

- Background noise assessment and velocity measurements on a cement grouting project in Ridgway, Colorado.(4)

Current Work on Grout Monitoring

The work has recently been extended to verify the source location computer code by using known sources to see how accurately they are predicted using the AE monitoring system. For an array as shown in Figure 2, with the actual source (a firecracker) at coordinates of (64", 61", 18") from an arbitrarily located point, the following results have been obtained.

DT1	DT2	DT3	DT4	DT5	RDC	ED	PA	ENG	RT	SLOPE	X	Y	Z
146	463	518	560	728	47	2527	28	62	236	1	59	65	21
148	469	504	545	730	26	1258	24	55	785	0	60	64	24
159	456	493	523	668	25	1120	26	56	566	0	59	61	20

where

DT = travel time in microseconds between sensor hits
RDC = ringdown count

ED   = event duration in microseconds
PA   = peak amplitude in millivolts
ENG   = energy
RT   = rise time in microseconds
SLOPE = descriptor of curve form
X,Y,Z = calculated values of coordinates

As seen in Figure 2, these predicted source values fall acceptably close to the actual source location.

Additional field testing was performed at Ridgway, Colorado on the cement grouting project previously mentioned. A number of new monitoring schemes were attempted. These involved sensor deployment methods in rock masses, shielding of sensors from background noise and different physical arrangements of the sensor array. Unfortunately, actual source location data was not obtained due to the high background noise with respect to the low AE event amplitudes. During this monitoring period the grout was an 8 (water) to 1 (cement) mix with injection pressures less then 10 psi. Thus we were dealing with very low level AE emissions.

Conclusions and Recommendations

The acoustic emission monitoring method offers to the geotechnical engineer a viable, economic and technically proved nondestructive testing method. In situations where a nonequilibrium phenomenon generates noises, the emissions can be sensed, transmitted and quantified in a reasonably straightforward manner. Soil instability, deformation, subsidence, etc., have been the thrust of most AE investigations to date. In this paper, however, a new application area of grout monitoring has been attempted. It differs from other work in that events, rather than counts, are the desired monitoring parameter. With multichannel pickup sensors these AE events can be used to locate the disturbance source. Used in conjunction with grouting, this disturbance is presumably the location of the grout front.

In the work described in this paper a series of laboratory and field activities were presented with the goal of developing a three-dimensional, real time graphic display of subsurface grouting activities. Each series was informative and progressively led to the following conclusions and recommendations for further research and development work.

- High background noise (e.g., construction equipment) and/or low amplitude AE signals limit the technique greatly.

- When monitoring can be conducted around these two limitations, the technique is indeed possible and very attractive.

- The initial computer codes (travel times and source coordinate calculations) are currently functioning.

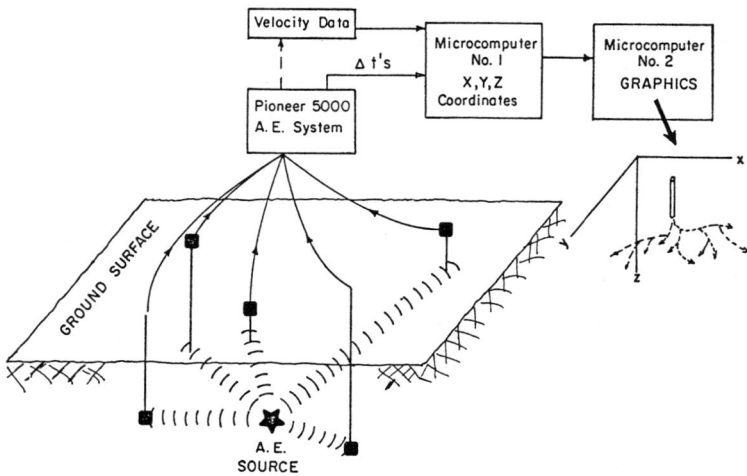

Fig. 1 - Schematic Diagram of Subsurface Disturbance, Travel to Transducer Pickups and Subsequent Analysis, Source Location, and Graphics Involved in this Proposed Project

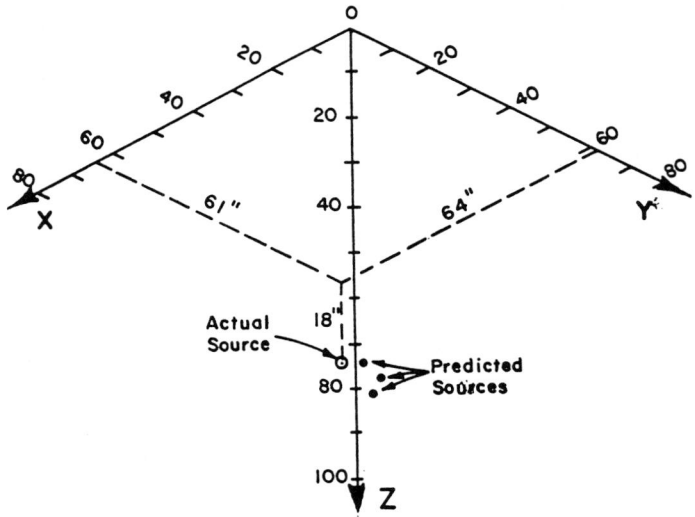

Fig. 2 - Three Dimensional Plot of Source Location

- The three-dimensional graphics code with storage capability is not a currently functioning program.

- Additional field work is necessary to debug the entire interactive system and to define its limitations.

- An interesting spin off of real time determination of in-situ velocity values (in any direction) was developed. This feature of the method has potential applicability in assessing the completeness of densification when using compaction grouting or deep dynamic compaction methods.

References

1. Hardy, H. R. Jr. and Leighton, F. W., Editors, Proceedings of Acoustic Emission/Microseismic Activity in Geologic Structures and Materials, Trans Tech Publ., 1977, 1980 and 1984.

2. Hardy, H. R. Jr., Applications of Acoustic Emission Techniques to Rock and Rock Structures, Acoustic Emissions in Geotechnical Engineering Practice, STP 750, V. P. Drnevich and R. E. Gray, Eds., ASTM, 1981, pp. 4-92.

3. Koerner, R. M., McCabe, W. M. and Lord, A. E., Jr., Acoustic Emission Behavior and Monitoring of Soils, Acoustic Emissions in Geotechnical Engineering Practice, STP 750, V. P. Drnevich and R. E. Gray Eds., ASTM, 1981, pp. 93-141.

4. Koerner, R. M., Leaird, J. D. and Welsh, J. P., Use of Acoustic Emissions as a Nondestructive Testing Method to Monitoring Grouting, Proc. Symp. on Innovative Materials and Techniques in Cement Gravity, Kansas City, Missouri, Sept. 1983, ACI, to appear 1984.

Acknowledgements

This work was performed under sponsorship of the U. S. National Science Foundation under Grant No. CEE81 - 08178 with Charles A. Babendreier as Program Director. This is under a joint University/Industry program with AET Corp. of Sacramento, California forming the industry coupling portion. GKN-Hayward Baker Company of Odenton, Maryland, cooperated by allowing us access to the field sites.

GROUT SLURRIES--THICK OR THIN?

Don U. Deere*, M.ASCE and Giovanni Lombardi**

ABSTRACT

Among the several issues in the grouting of dam foundations that remain in discussion is the desirable thickness, or viscosity, of the grout slurry. Proponents of the thin mix believe that finer cracks can be grouted and better penetration achieved by using a thin mix. Proponents of the thicker mix believe that a denser, stronger, more uniform, and chemically resistant hardened grout results from using the thicker mix, and that with sufficiently high pressures and length of time fine cracks can also be penetrated. In this paper some recent experiences in dam grouting in several countries are discussed where bentonite or Intraplast additives were used to control sedimentation, viscosity, grout density, set time, and strength. Also presented are recently developed theoretical and laboratory studies regarding the viscosity of grout slurries which point out the role of cohesion (yield point) of the grout in determining the penetration and uplift pressure exerted by the injected slurry. Results would appear to favor thicker, stable mixes.

INTRODUCTION

Debates about the relative advantages of thin versus thick grouts have been common since the development of a literature on grouting. The advocates for thin grouts obviously feel that only with very thin, watery grouts such as 9:1 or 6:1 water:cement by volume (6:1 or 4:1 by weight) can the narrow cracks in a rock mass be penetrated. They often state that the entrance to a thin crack can be blocked off prematurely by a thick grout slurry.

Advocates for thicker grouts such as 2:1, 1.5:1, or 1:1 water: cement by volume (1.3:1, 1:1, or 0.67:1 by weight) with or without a small percentage of admixture usually are proponents for the use of a "stable" mix, that is, one in which settling out of the cement grains is minimal. A sedimentation of 5 percent or less of the cement in 2 hours (volume of clear water at the top of a 1000-ml graduated cylinder divided by the original grout volume) is a value commonly chosen to characterize a stable mix.

The thick-mix proponents believe that only with a stable mix that maintains its flow properties can one get optimal filling and at the

---

*Consultant, 6834 S.W. 35 Way, Gainesville, Florida, 32608 and Adjunct Professor, Dept. of Civil Engineering, University of Florida.

**Consultant, Studio Lombardi, Via Ciseri 3, CH-6600, Locarno, Switzerland.

GROUT SLURRIES

same time guarantee a strong, dense hardened grout resistant both to physical and chemical deterioration. Both Deere (1) and Houlsby (2) among others have noted the advantages of a stable mix.

RECENT EXPERIENCES

At the Itaipu dam recently completed on the Parana river between Brazil and Paraguay the grouting was done almost entirely with a single mix, 1:1 W:C by weight with 1% to 2% of bentonite (depending on the liquid limit of the bentonite). The Marsh funnel viscosity was from 38 to 40 sec., the sedimentation about 5%, and the slurry specific gravity 1.5.

At the Alicura dam recently completed on the Limay river in Argentina the first part of the grouting program was done with a mix of 1:1 W:C by weight with 2% bentonite (by weight of cement). This gave similar properties: Marsh value of 35 to 38 sec., sedimentation of 3 to 5%, and a slurry specific gravity of 1.50. The mix was later changed to a denser mix of 0.67:1 W:C by weight but with 1% Sika Intraplast (instead of 2% bentonite). This lowered the Marsh value to 32 sec. while increasing the 28-day strength over that with the bentonite admixture.

An additional advantage of using Intraplast is that it was added directly to the mixer and did not have to be separately mixed, hydrated, and stored in a separate agitator as for the bentonite.

In New Zealand at the Clyde dam under construction on the Clutha river a series of laboratory and field tests were made which allowed an excellent mix and procedure to be developed. They use a 1:1 mix, W:C by weight (1.5:1 by volume) with 0.5% bentonite. The bentonite is pre-mixed in the regular mixer for 5 minutes with the water; the cement is then added for an additional 3 minutes of mixing. This procedure allows for some of the hydration of the bentonite to take place, sufficient to reduce sedimentation to less than 2%, but apparently not too much, so that the Marsh value is only 32-34 sec. The specific gravity of the slurry is 1.50-1.52.

A similar procedure for partial hydration of the bentonite was used in the Rocky Mountain Project, Georgia Power Company, for grouting the pressure tunnel lining and at another current project in Washington for the dam grouting.

In the following sections the results of a number of recent theoretical studies are presented. These together with the recent field experiences may be helpful in studying the question--thick or thin?

SOME THEORETICAL CONSIDERATIONS ON GROUTING

New Investigations

Previous theoretical studies of cement grouting usually have considered the slurry to be similar to water, that is, a Newtonian fluid, but with a somewhat higher viscosity. In fact, a stable grout slurry is not a Newtonian fluid but a visco-plastic Bingham fluid which possesses both viscosity and cohesion (yield point).

The <u>viscosity component</u> determines the rate at which a grout travels away from the grout hole under a given pressure and for a given thickness of an open joint. It is the <u>cohesion component</u>, however, that determines the final distance of penetration. The travel distance is always finite. The cohesion of the grout, thus, has the function to limit the extension of the grouted zone while the viscosity component influences the time needed to fulfill the grouting.

In a recent paper Lombardi (3) developed the equations which allow the maximum radius of penetration (Rmax), the maximum volume of injected grout (Vmax), and the maximum total uplift force (Fmax) to be computed. These are as follows:

$$R_{max} = \frac{p_{max} \cdot t}{C} \quad (1)$$

$$V_{max} = \frac{2\pi p^2_{max} \cdot t^3}{C^2} \quad (2)$$

$$F_{max} = \frac{\pi p^3_{max} \cdot t^2}{3C^2} \quad (3)$$

where p_{max} is the final applied pressure,
t is the half-thickness of the joint, and
C is the cohesion (yield point) of the grout slurry

A computer program has also been developed (3) to allow the transient values of R, V, and F to be obtained at any point in time during the grouting process.

For determining the values of the viscosity and cohesion of the grout a co-axial viscometer may be used. However, a simpler device has been developed for determining the cohesion, the plate cohesionmeter (3). This is a metal, rectangular plate with roughened surfaces which is lowered into the grout slurry. It is withdrawn and the slurry is allowed to drip from the plate. The amount of slurry held by the roughened surfaces is determined by the increase in weight of the plate. This allows the cohesion to be calculated.

The Marsh funnel is then used to get the "apparent" viscosity of the slurry. From the apparent viscosity and the cohesion, one may determine the true viscosity from prepared graphs (3).

Some Parametric Studies

It is of interest to use the developed equations and computer programs to study the influence of the width of the open joint and the consistency of the grout (thin, medium, or thick) on the radius of grout penetration, the flow rate of the grout, the volume of grout injected, and the uplift pressure.

Two crack widths are investigated, a fine crack of 0.5 mm (0.02 in.), and a medium fine crack of 1 mm (0.04 in., or a little greater than 1/32 in.). Three grout slurries are considered: a thin slurry (30 sec. Marsh flow time), a medium slurry (40 sec. Marsh flow time), and a thick slurry (50 sec. Marsh flow time). These have approximately the following values of cohesion and viscosity, normalized by dividing each by the specific weight: thin slurry (C=0.20 mm; viscosity=0.68 $\times 10^{-6}$MS); medium slurry (C=0.75 mm; viscosity=1.30$\times 10^{-6}$MS); and thick slurry (C=1.05 mm; viscosity=1.85$\times 10^{-6}$MS). It will be noted that the viscosity of the thick slurry is about 3 times that of the thin one but that the cohesion is 5 times greater. Some results are also presented for water (cohesion=0; normalized viscosity=0.13$\times 10^{-6}$MS). The grouting simulation assumes a large grout pump with elliptical characteristics [max. discharge 100 l/s (1580 gal/min); max. pressure 2000 kPa (20 kg/cm^2 or 284 lb/in^2)].

Radius of Grout Penetration.

Pertinent data are presented in Table 1. It is noted that for the ½-mm crack at the end of 15 minutes the water has penetrated a radial distance of 55 m, the thin mix a distance of 28 m, and the thick mix only 13 m. It takes about 4 hrs (or 16 times the 15-min. period) for the travel distance to approximately double (Note: Calculations were not made beyond distances of 100 m). For the medium and thick mixes, little is gained in penetration by grouting beyond 4 hrs. The thin mix is apparently still moving out slowly while the water flow would continue indefinitely since it has no cohesion.

Table 1. Radius of Grout Penetration as Function of Crack Width and Grout Consistency

(a) For ½-mm Crack Width

Radius of Penetration, m:	Water	Grout Slurry Consistency		
		Thin	Medium	Thick
t=15 min	55	28	17	13
t= 1 hr	100+	49	26	20
t= 4 hr		81	34	24
t= 8 hr		100	36	25

(b) For 1-mm Crack Width

t=15 min	60	43	32	25
t= 1 hr	100+	80	50	38
t= 4 hr		100+	67	48
t= 8 hr			72	50

Note: 25.4 mm=1 inch   1 m=3.28 ft   t=time

For the 1-mm crack width, the time to achieve similar penetration is about double that required for the ½-mm crack. It is apparent that more than 8 hrs would be necessary to achieve full penetration for the thin and medium mixes.

Grout Flow Rate. Pertinent data are presented in Table 2. It is noted that at the end of 15 minutes for the ½-mm crack a flow rate of 312 liters/min is being achieved while for the thin grout it is only 69 liters/min, and for the thick grout is is only 12 liters/min. It is also obvious from Table 2 that little is being gained for the medium and thick mixes by grouting beyond 4 hrs.

Table 2. Grout Flow Rate as Function of Crack Width and Grout Consistency

(a) For ½-mm Crack Width

		Grout Slurry Consistency		
Flow Rate, lts/min	Water	Thin	Medium	Thick
t=15 min	312	69	22	12
t= 1 hr	307	50	9	4
t= 4 hr		28	2	0.6
t= 8 hr		18	0.6	0

(b) For 1-mm Crack Width

	Water	Thin	Medium	Thick
t=15 min	760	359	156	93
t= 1 hr	756	297	76	36
t= 4 hr			17	6
t= 8 hr			6	2

Note: 25.4 mm=1 inch   1 lt/min=0.26 gallons/min   t=time

For the 1-mm crack width as compared to the ½-mm crack width, it is seen that the flow rate at the end of 15 min of grouting is more than twice as great for water, 6 times greater for the thin mix, and 8 times greater for the thick mix. The thick mix obviously is penetrating the wider crack much easier than it did the finer crack.

Volume of Grout Slurry Injected. Table 3 indicates the same trends as the previous tables. By examining the numbers, the time beyond which little is being gained can be visualized.

It is interesting to note that a similar volume of thin grout can be injected in 15 min. as the thick grout in 8 hrs. (½-mm crack) or about 2 hrs. (in 1-mm crack).

Uplift Force Exerted. The uplift forces exerted are given in metric tons in Table 4. The enormity of these forces is striking, particularly for the thinner mixes. The forces have been obtained by integrating the pressure in the joint from its maximum value at the grout hole to near zero at its leading edge of penetration giving a nearly triangular distribution. Therefore, with thinner mixes and with increasing time the area of injection increases as does the total uplift force.

Table 3. Volume of Grout Slurry as Function of Crack Width and Grout Consistency

(a) For ½-mm Crack Width

Vol. of Grout Injected, m^3	Water	Grout Slurry Consistency		
		Thin	Medium	Thick
t=15 min	4.8	1.2	0.5	0.3
t= 1 hr	15.9	3.7	1.1	0.6
t= 4 hr		10.3	1.8	0.9
t= 8 hr		15.6	2.1	1.0

(b) For 1-mm Crack Width

	Water	Thin	Medium	Thick
t=15 min	11.4	5.8	3.1	2.0
t= 1 hr	31.4	20.2	7.7	4.5
t= 4 hr			14.1	7.1
t= 8 hr			16.4	7.9

Note: 25.4 mm=1 inch   1 m^3=35.3 ft^3 t=time

Table 4. Uplift Force as Function of Crack Width and Grout Consistency

(a) For ½-mm Crack Width

Uplift Force, Metric Tons:	Water	Grout Consistency		
		Thin	Medium	Thick
t=15 min	48,000	61,000	39,000	26,000
t= 1 hr	158,000	255,000	106,000	68,000
t= 4 hr		942,000	225,000	115,000
t= 8 hr		1,670,000	267,000	129,000

(b) For 1-mm Crack Width

	Water	Thin	Medium	Thick
t=15 min	18,000	109,000	123,000	88,000
t= 1 hr	48,000	551,000	401,000	252,000
t= 4 hr			633,000	359,000
t= 8 hr			1,055,000	514,000

Note: 25.4 mm=1 inch   1 metric ton=1000 kg=2205 lbs.   t=time

While the table indicates that water gives the lowest uplift pressures, this trend would reverse with time because the water penetration continues with time. Therefore, the area affected and the uplift force would also increase indefinitely. Moreover, if the water penetration were to be stopped by a barrier or by plugging, its triangular distribution would become rectangular and would give the largest unit uplift pressure and total uplift force.

Maximum Radius of Penetration. Table 5 shows the maximum radius of penetration for the different grouts and for the two crack widths. The cohesion of the thick slurry causes plugging of the ½-mm crack at a radial distance of 27 m and for the 1-mm crack at 53 m. From equation (1) it is seen that the maximum radius of penetration is directly proportional to the crack width and inversely proportional to the cohesion of the grout. Therefore, to prevent large losses of grout into cracks greater than 0.5-mm thickness, both low pressures and thick to very thick mixes with appreciable cohesion should be used.

Table 5. Maximum Radius of Penetration as Function of Crack Width and Grout Consistency

	0.5-mm Crack	1-mm Crack
Thick Mix	27 m	53 m
Med. Mix	39	78
Thin Mix	156	313

Effect of Additives

Preliminary results of the recent studies of the viscosity and cohesion of grouts with and without additives have shown:
--Decreasing the water:cement ratio increases both the viscosity and the cohesion, but the cohesion proportionally more.
--Adding bentonite increases both the viscosity and cohesion, but the cohesion proportionally more.
--Adding a fluidifier (Intraplast, Intracrete, or Rheobuild) decreases the viscosity, and probably to a lesser extent the cohesion.
--For the same Marsh flow value, a grout with fluidifier will be denser and have a greater 28-day compressive strength than one with bentonite.

Bentonite should therefore be used only to increase the cohesion and limit the travel distance (it is just the contrary of lubrication, as sometimes claimed). A fluidifier should be used to ease the penetration into thin cracks and to increase the travel.

DISCUSSION

Dense, stable mixes are desirable to achieve the following to the greatest extent possible:
--Complete filling of the voids and joints.
--High mechanical strength.
--Good bond to rock.
--Resistance against chemical leaking.
--Predictability of the grouting process.
--Limited travel of the grout avoiding useless losses.
--Reduced risk of uplift.

The main results of the theoretical computations are:
--The travel distance is always limited.
--It depends on cohesion, pressure, and crack width.
--While the theoretical travel is never completely reached, as it would require infinite time, the rate of grout injection decreases drastically as the theoretical distance is approached and grouting may be stopped.

--The uplift or splitting forces are limited, being less for the thicker mixes. They are functions of pressure and radius of grout penetration.

While it is realized that the theoretical values will differ considerably from those which will actually be achieved in the field because of the greater complexity of the joint geometry and surface characteristics, the general trends will still apply and may be used in formulating a grouting philosophy.

PROPOSED GROUTING SCHEME

To take advantage of the previously noted theoretical indications and practical requirements the following scheme of hole layout and grouting sequence and pressures is proposed. It is based on the principle of using the primary and secondary holes for grouting the larger cracks and the tertiary and quarternary for grouting the finer cracks. For this case a single-line curtain and a single stable grout slurry of about Marsh flow time of 40-50 sec is considered.

1--Primary holes on 20-m spacing with moderate pressure of p.
2--Secondary holes split-spaced to 10-m at pressure of 1.5 p.
3--Tertiary holes split-spaced to 5-m between the P and S holes at pressure of 2 p.
4--Quarternary holes at pressure of 2.5 p between previously grouted holes but not split-spaced, rather 3 m from the T holes but 2 m from the P and S holes.

This scheme may also be adapted to use the same grouting pressure for all holes but a thicker mix (such as 60-sec. Marsh) for the P and S holes and thinner mixes for the T and Q holes. Of course, the two procedures may be combined in which a thick mix and a moderate pressure are used on the P holes, followed progressively by somewhat higher pressures and thinner mixes on the S, T, and Q holes. A thick stable mix will not choke off a fine crack if it is large enough to be grouted at all; it will just not penetrate very far, either because of time restraint or because the cohesion causes plugging after some finite penetration.

This scheme may also be even more effectively applied to a multi-line curtain, reserving the higher pressures and/or thinner mixes for the center line of holes.

ACKNOWLEDGEMENT

The writers wish to thank Dr. G. Anastasi of Studio Lombardi for his helpful comments and for his development of the computer program and the performance of the laboratory tests.

APPENDIX.--REFERENCES

1. Deere, D.U. "Cement-Bentonite Grouting," Grouting in Geotechnical Engineering, Proceedings of Conference sponsored by the Geotechnical Engineering Division, ASCE, New Orleans, 1982, pp. 279-300.

2. Houlsby, A.C. "Optimum Water:Cement Ratios for Rock Grouting," Grouting in Geotechnical Engineering, Proceedings of Conference sponsored by the Geotechnical Engineering Division, ASCE, New Orleans, 1982, pp. 317-331.

3. Lombardi, G. "The Role of Cohesion in Cement Grouting of Rock," Paper presented to the Fifteenth International Congress on Large Dams, Lausanne, Switzerland, 1985 (in press).

SUBJECT INDEX

Page number refers to first page of paper.

Acoustic detection, 149
Acrylate polymer grout, 1
Asphalts, 76

Cement grouts, 34, 132
Chemical grouting, 92
Chemical grouts, 1
Compaction, 104
Computer applications, 123, 132

Dam foundations, 76, 123
Dams, earth, 92
Dams, embankment, 104
Densification, 104

Foundations, 104

Grain size, 27
Grout, 27, 156
Grout curtains, 92
Grouting, 104, 123, 149

Injection, 149

Leakage, 76

Monitoring, 123, 132, 149

Nondestructive tests, 149

Quality control, 123, 132

Sand, 1
Silicates, 1
Slurries, 156

Testing, 1

Viscosity, 27, 156

Water content, 34

AUTHOR INDEX

Page number refers to first page of paper.

Baker, Wallace Hayward, 104

Deans, John, 76
Deere, Don U., 156
Demming, Michael, 123

Faught, Kenneth L., 92

Graf, Edward D., 92
Grant, Leland F., 132

Houlsby, A. Clive, 34

Karol, R. H., 27
Koerner, Robert M., 149

Krizek, Raymond J., 1

Leaird, James D., 149
Lombardi, Giovanni, 156
Lukajic, Boro, 76

Madden, Mark, 1

Rhoades, Daniel J., 92
Rogers, James L., 123

Sands, Richard N., 149
Smith, Grant, 76

Tula, Alex, 123

9774

SOUTHEASTERN MASSACHUSETTS UNIVERSITY
TC540.I78 1985
Issues in dam grouting

3 2922 00010 323 1

DATE DUE

ILL: 2483588		
DUE: 3-21-93		
APR 02 1993		
ILL 6699192		
Due 3/16/95		
MAR 23 1995		
ILL# 2228205		
Due 01/06/03		
DEC 13 2002		
261-2500		Printed in USA